SpringerBriefs in Geography

SpringerBriefs in Geography presents concise summaries of cutting-edge research and practical applications across the fields of physical, environmental and human geography. It publishes compact refereed monographs under the editorial supervision of an international advisory board with the aim to publish 8 to 12 weeks after acceptance. Volumes are compact, 50 to 125 pages, with a clear focus. The series covers a range of content from professional to academic such as: timely reports of state-of-the art analytical techniques, bridges between new research results, snapshots of hot and/or emerging topics, elaborated thesis, literature reviews, and in-depth case studies.

The scope of the series spans the entire field of geography, with a view to significantly advance research. The character of the series is international and multidisciplinary and will include research areas such as: GIS/cartography, remote sensing, geographical education, geospatial analysis, techniques and modeling, landscape/regional and urban planning, economic geography, housing and the built environment, and quantitative geography. Volumes in this series may analyze past, present and/or future trends, as well as their determinants and consequences. Both solicited and unsolicited manuscripts are considered for publication in this series.

SpringerBriefs in Geography will be of interest to a wide range of individuals with interests in physical, environmental and human geography as well as for researchers from allied disciplines.

More information about this series at http://www.springer.com/series/10050

Mary J. Thornbush · Oleg Golubchikov

Sustainable Urbanism in Digital Transitions

From Low Carbon to Smart Sustainable Cities

 Springer

Mary J. Thornbush
Faculty of Environmental Studies
York University
Toronto, ON, Canada

Oleg Golubchikov
School of Geography and Planning
Cardiff University
Cardiff, Wales, UK

ISSN 2211-4165 ISSN 2211-4173 (electronic)
SpringerBriefs in Geography
ISBN 978-3-030-25946-4 ISBN 978-3-030-25947-1 (eBook)
https://doi.org/10.1007/978-3-030-25947-1

This Springer imprint is published by the registered company Springer Nature Switzerland AG
The registered company address is: Gewerbestrasse 11, 6330 Cham, Switzerland

Preface

The idea of urban sustainability has experienced important transformations in the past decades. The emphasis on eco-cities and the low carbon agenda due to climatic challenges is increasingly combined with the rise of smart cities. Sustainable urbanism, thus, re-emerges in an upscaled fashion to engulf smart cities and innovative technical solutions embracing information and communication technology or ICT. The initial driver was to work towards service and resource use efficiency at a smaller scale, but the concept of the smart city has developed from this original ambition to one that applies to entire cities and urban areas, and no longer just the transportation system or buildings. Modern applications involve increasingly greater connectivity or integration, with the involvement of multiple stakeholders and city components. The smart city is based on automation and monitoring by sensors and Big Data collection, which are used to improve performance and to inform governance. This transformation, however, raises new critical questions, including whether smart sustainable cities become too technocratic in actual operation, but also with regard to citizen involvement in such a technologically automated environment. Moreover, problems that are associated with cybersecurity and the use of Big Data, including personal privacy—and ultimately democracy—need to be addressed. This brief reviews these important contemporary concerns. It also discusses the degree to which smart cities function to improve the quality of life for urban citizens and their role in enacting the 'simple life' concept for sustainable urban development.

Acknowledgements We are grateful for the feedback received from and discussion stimulated during a presentation on smart cities, energy, and urban strategies by the authors as part of the Cities Research Centre at Cardiff University, Wales, held on 24 April 2018. We are also very grateful to an anonymous engineer and computer scientist who reviewed our manuscript.

Toronto, ON, Canada   Mary J. Thornbush
Cardiff, Wales, UK   Oleg Golubchikov

Contents

About the Authors

Dr. Mary J. Thornbush is presently Researcher in the Ecological Footprint Initiative based in the Faculty of Environmental Studies at York University, Canada. She has over 80 publications in the areas of applied geomorphology and environmental and urban sustainability. Her doctoral thesis at the University of Oxford addressed urban sustainability through a study of air emissions from transport in central Oxford and investigated their impacts on the weathering of its historical limestone buildings. Her relevant publications include a special section on Geography, Urban Geomorphology and Sustainability in the journal *Area* (2015) as well as books such as *Vehicular Air Pollution and Urban Sustainability: An Assessment from Central Oxford, UK* (2015, Springer) and a volume on *Urban Geomorphology: Landforms and Processes in Cities* (2018, Elsevier) in the area of sustainability research.

Dr. Oleg Golubchikov is Reader in Human Geography at the School of Geography and Planning at Cardiff University. He previously worked as an academic at the Universities of Oxford and Birmingham. He has also held visiting academic positions in Sweden, Finland, and Russia. His research interests lie with urban political geography, sustainable cities, and energy geography. His recent research interrogates the relationships between spatial governance and urban and regional transformations in the context of major contemporary societal 'projects' including post-socialist and post-carbon transitions. He has developed research projects and collaborations across Europe and in the BRIC countries. His research also informs international policies. He has advised the United Nations on aspects of sustainable housing, urban development, and low carbon cities.

Chapter 1
Introduction

Abstract A literature search was performed to track the development of the concept of smart cities as it appears in known published works. Google Scholar was the chosen search engine as representative of a comprehensive database. Based on this search, the chapter highlights trends in smart city development, beginning at the building scale and working upwards to city, regional, and ultimately national levels. European examples demonstrate how cities have upheld smart development to convey the potential for expansion as well as upscaling and multi-scaling. In addition, rebranding is considered alongside upscaling.

Keywords Google Scholar · Internet searches · Existing smart cities · Sustainable cities · Smart energy cities · Building scale · City scale · Multi-scalar · Smart development · Energy framework

Recent advancements in urban sustainability have embraced technology to foster efficiency and organisation. In the context of increasing literacy for the use and consumption of hi-tech goods, cities have adopted technological systems in the form of integrated information and communication technology (or ICT) like never before. These systems have become 'intelligent' through developed sensing capabilities and automation via computer programming, with devices connected (in real time) by way of the Internet in tiered platforms. In this way, it has become possible for companies and governments to collect information and generate Big Data on almost every facet of urban life.

Indeed, the modern reach of technology in cities is impressive. Devices linked to systems are deployed to track various types of information, generating varied and massive databases, and the Big Data that is being compiled has the potential to function in a variety of ways and fill different niches. It is difficult to pinpoint exactly where this began, but arguably the placement of surveillance cameras as a security measure in cities, such as CCTV systems, are what triggered the application of smart technology in the growing metropolis. Another consideration that will form the discussion in this brief is that of the origin of smart cities in sustainable cities more specifically, beginning with energy-efficient buildings and cities (Chap. 3) to the smart energy cities of today (addressed in Chap. 6).

M. J. Thornbush and O. Golubchikov, *Sustainable Urbanism in Digital Transitions*, SpringerBriefs in Geography, https://doi.org/10.1007/978-3-030-25947-1_1

Recent developments in this topical area merit attention to the concept and its practical application in society. This brief is a timely contribution to this, with its focus on the evolution of the concept deploying exemplar real-world examples, as 'actually existing' smart cities, based on targeted city strategies and detailed cases. The overarching purpose of the first portion of this brief is to present the smart city by tracking its origins and path of development.

1.1   The Rise of Smart Cities

Published papers broadly on smart cities were accessed through Google Scholar as a comprehensive database on academic and related literature, including various databases and sources of information that are simultaneously easily accessible world-wide and can be used to verify the findings on which the ensuing discussion is based. To narrow the search, a variety of search word combinations were used, excluding patents and citations (Table 1.1; first accessed on 27 October 2017). For example, 'smart cities' resulted in a broader search than 'smart city' and this was narrowed further using 'existing smart city' and especially through the deployment of 'actually existing smart city' searches (see Table 1.1).

The purpose here was to examine the mainstream literature, accessed by way of Google Scholar, in order to identify relevant thematic components. Also, as presented later, the intent was to also compile a roster of actually existing smart cities on which to base an analysis of real-world cases for support.

There were 30 varieties of search strings that could have been deployed for literature acquisition in this metareview. The most limiting word string was 'actually existing smart cities', which produced the lowest number of results (see Table 1.1). The overall range was between 'smart cities' and 'actually existing smart cities', with the greatest reduction in the results when 'existing smart city' was deployed. For this reason, this search algorithm 'existing smart city' was used to encompass almost 300 results in Google Scholar. The purpose of doing so was to use search results that provided a middling result that was not too broad or focussed so that thematic categories could be derived from the search results. The resulting discussion is based on these thematic findings and develops them in a consideration of broader issues in the smart cities' literature, again with a focus on existing applications of concepts and ideas contained within the literature rather than a utopian view of what the smart (energy) city should ideally entail. By comparison, databases such as GEOBASE (Scopus) found 20% of what was located using a Google search (Fig. 1.1).

These findings are similar to results by Pollio (2016a, Fig. 1, p. 518), who used Google to search for 'smart city' and discovered peaks in queries (in the research index for the term) in 2012 and 2013. According to published works, the trend is one of non-linear (exponential) increase since 2009, with a growth since 1997 of 50% up until 2017 (see Fig. 1.1). These findings accord with the interpretation by Pollio (2016a) for Italy of the popularisation of the smart city for countries experiencing austerity measures to combat the economic crisis (which happened globally in the

Table 1.1 Effects of search words and strings on Google Scholar page results

Search word(s)	Results
'smart cities'	51,900
'smart city'	42,500
'existing smart city'	302
'existing smart city' social	284
'existing smart city' planning	278
'existing smart city' governance	233
'existing smart city' urban governance	226
'existing smart city' energy	222
'existing smart city' geography	198
'actually existing smart city'	150
'actually existing smart city' social	147
'actually existing smart city' planning	144
'actually existing smart city' governance	134
'actually existing smart city' urban governance	131
'actually existing smart city' sustainability	119
'existing smart city' democracy	117
'actually existing smart city' geography	116
'actually existing smart city' energy	99
'existing smart city' 'urban governance'	89
'existing smart cities'	81
'actually existing smart city' democracy	80
'existing smart city' social justice	74
'existing smart city' justice	74
'existing smart city' neoliberalism	67
'actually existing smart city' neoliberalism	63
'actually existing smart city' justice	61
'actually existing smart city' social justice	58
'existing smart city' 'social justice'	27
'actually existing smart city' 'social justice'	20
'actually existing smart cities'	18

aftermath of the 2008 global financial crunch) as for Barcelona, Spain (March & Ribera-Fumaz 2016). The authors argue that the implementation of the smart city at a time of economic crisis was to ground utopias.

Consequently, it can be argued that the popularisation of the smart city occurred at a time when the world was experiencing economic crisis in 2008, with Google search engines capturing public (Pollio 2016a) and scholastic attention turning to the smart city at a time of economic downturn, as 'political technologies' were deployed to leverage neoliberal agendas calling for 'pro-innovation' public spending. Arguably,

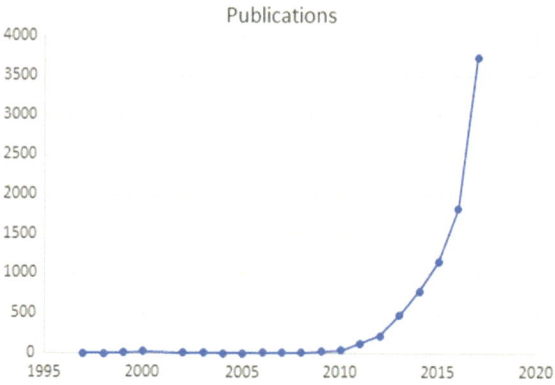

Fig. 1.1 Trends in publications for 'existing smart cities' in GEOBASE (Scopus)

people turned to technology and innovation to fuel their business (including the rampage of start-ups that are still prevalent in response to high unemployment and job cuts due to austerity measures in what can be considered to be a degrowth economy). So, it is more than the promotion of a techno-utopian view that is being supported here; as espoused by Pollio (2016b), the smart city was an effective rebranding of the urban in the creation of techno-utopian imageries.

The conceptual development of the smart city with regard to urban sustainability since the early 2000s has much been focused on the building scale and smart buildings, particularly in interplay with energy efficiency (Fig. 1.2) and other small-scale 'niches' such as transportation systems and was, subsequently, upscaled to city level after 2008. This suggests that the root of the concept was already active (at some scale) before the onset of the global economic crisis and was propelled thenceforth to the present day. As an initial focus, the auspice of 'low carbon futures' acted as a mechanism from which it emerged (Fig. 1.3), being supported by climate change policy founded around 2007. It has been rooted as 'urban resilience' since 2011, which became a pronounced development particularly in 2013–2015. Since the mid- to late 2000s, smart cities have shifted in alignment with austerity measures during the debt crisis, creating a market for technology, especially in the domain of computing in the current Digital Age. The most recent development since the early 2010s has been for the establishment of 'smart energy cities'; this will be the topic of Chap. 6.

1.2   Upscaling to the Smart City

While smart cities as part of urban sustainability have captured the attention of researchers since the end of the 2000s, the plethora of literature is especially evident in the academic literature around 2015–2016, with a growth of 50% in this timeframe (see Fig. 1.1). An initial focus on particular issues such as energy efficiency in buildings was upscaled to the city level since after 2008—as for example evident

Fig. 1.2 Historic evaluation of the smart city concept from energy efficiency to digital transformation

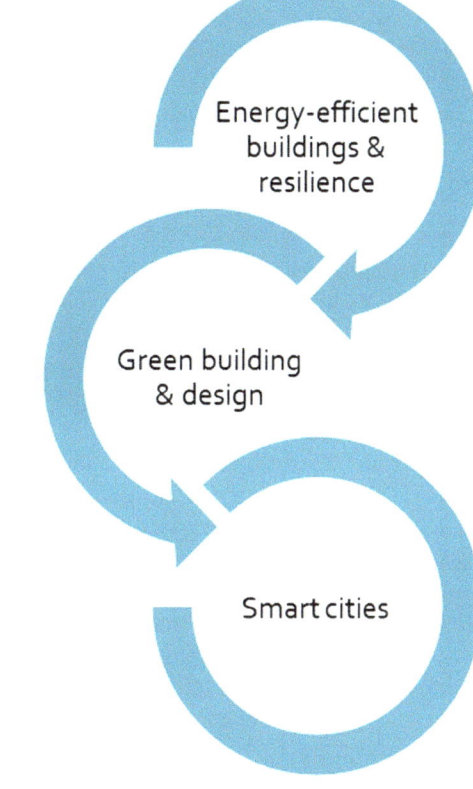

Fig. 1.3 Emergence of smart energy cities from low carbon urbanism and energy-efficient cities

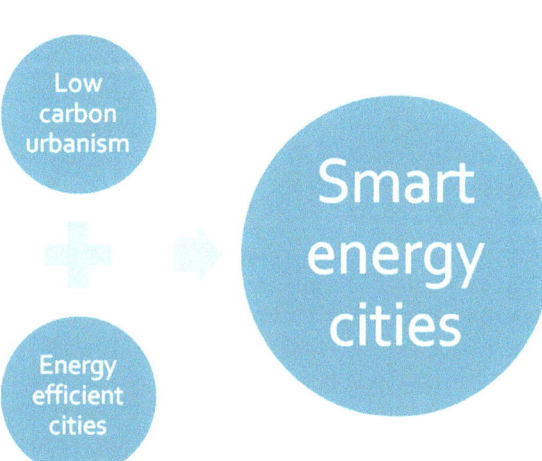

for Milano, Italy in preparation for Expo 2015 (Causone et al. 2017). This is based on a focus on energy that stems from low carbon futures, with policy support from climate change concerns. There was, consequently, also a turn to urban resilience—becoming urban sustainability research, especially in 2013–2015. The smart cities shift is evident since the mid-2000s—aligning with austerity measures during the debt crisis, which will be conveyed later in Chap. 5, with urban consultancy by companies creating a market for computer platforms in the Digital Age.

This has constituted a 'piecemeal' development of the smart city, with individual buildings being investigated first in energy journals (e.g., Leiria Polytechnic Institute, Portugal, where photovoltaic panels were planned for emplacement; Galvão et al. 2017), accompanied by renewable energy to feed grids and electric vehicles contributing to infrastructure and mobility (e.g., Agudo-Peregrina & Navío-Marco 2016; also, Mendoza et al. 2015, who advocate the provision of clean energy supplied by pergolas to e-bikes) to support smart development. Consequently, smart urbanism spread to encompass entire cities and regions, as portrayed in Chap. 5 for Barcelona, Spain, which has been reimaged as a smart and self-sufficient city (March & Ribera-Fumaz 2016). According to the authors, this has established the 'multi-scalar' city based on public-private partnerships upholding a distributed network of collaborations in Barcelona, occurring at the local scale (as local projects) to international collaborations as part of international projects as well as governance arrangements that range from the local to international scales.

The situation with cities such as Barcelona and Milano, and others, is one of rebranding rather than simple upscaling from the building to city and regional levels. This reimaging of cities as 'smart' is arguably linked with an entrepreneurial approach (outlined in Chap. 5) that has been associated with austerity measures in Italy (Pollio 2016a). The latter has relayed to the entire country's policy rather than individual cities, indicating a spread of smart development in Italy in response to annihilated fiscal budgets (Pollio 2016a). This is part of a technological solution (also discussed in Chap. 5) that is at the core of smart development. In Europe at least, this may be connected with the EU-Innovation Union in combination with European Digital Agenda, fostering the vision of Europe 2020 as a strategy for smart, sustainable, and inclusive growth.

This development begs the question whether rebranding or upscaling is occurring in cities more broadly around the world. It seems that while some cities, such as Barcelona, are being reimaged and rebranded based on a fundamental entrepreneurial approach to smart development, there is also evidence of upscaling across countries like Spain, where 62 cities have been investigated (Aletà et al. 2017). Smart development has also been popular in Italy and Greece as an economic mechanism driving entrepreneurship (see Chap. 5). Generally, however, the trend has been one of establishing energy-efficient buildings—upscaled up to the city level—to foster resilience in urban development (see Chap. 3) and promoting green building and design before progressing to the smart city. Although this brief focuses on this aspect of smart development, it is noteworthy that other elements of smart cities, including transport, security, public services, and so on have also been vehicles for their propagation. The overarching aim has been that by incorporating ICT, it is possible to improve the quality of life for urban citizens through the enhanced quality and better performance of urban services, such as transport, utilities, energy, and so on, while reducing resource consumption, waste, and costs. As evidenced in the USA, for instance, smart cities have been promoted through transportation—with the Smart City Challenge, for example, promoting smart growth and development in North American cities.

The current piecemeal development that works towards realising smart cities around the world does not preclude an all-round smart development that envelopes the globe (at an international scale) and is no longer occurring only at the city-to-regional scale. Nevertheless, cities are driving the transition, although ultimately it will no longer just be the smart city that encompasses broader smart development (e.g., transboundary flows, affecting local livelihoods, health, and well-being, from the home to neighbourhood and city to the region/nation and, having transboundary impacts on environment, well-being, and climate change, to finally encompass the globe, cf. Ramaswami et al. 2016) evident at the national level, as already apparent in some European countries, such as Italy and Spain, seeking to represent 'smart nations'. This will be developed in the next chapter that presents a roster based on strategies for smart cities compiled for this research and also including some detailed case studies to represent growing smart initiatives from developing and already developed parts of the world.

1.3 Brief Aims

Regardless of their beginnings, smart cities have come a long way in contemporary urban environments. They continue to advance as new technologies and systems integration are achieved, driven by innovation and fed by entrepreneurialism. Their diversity evokes an interdisciplinary approach to smart cities that brings together academics and practitioners as well as researchers from a plethora of disciplines. This brief will further deliberate the origin and evolution of smart cities and their rationale as well as issues of development, focusing on different strategies adopted to realise them as well as the potential challenges.

The rationale for this brief stems for a need to engage with smart cities in a way to allow for critical assessment of their development, as regarding how smart cities can be disruptive to the long-term course of sustainable cities. There are tensions within social-technical systems needing critical address. This needs to be done from an understanding that acknowledges the diversity in actually existing cases, representing different priorities in future development. This needs to be executed based on an empirical approach that is grounded in real-world cases, as provided by smart city strategies denoting actually existing smart cities. Finally, by recognising diversity, it is possible to grasp issues affecting convergence and divergence in smart development and be able to capture the digital divide.

The main conceptual inputs of this research will be relayed in the next couple of chapters as part of an investigation into the literature, tracking the concept and its evolution. This will be followed by another two chapters making different methodological inputs to the study, such as a roster of city-scale strategies (Chap. 4) used to identify smart cities around the world and detailed case studies (see Chap. 4) chosen from developing and developed countries. Challenges will be considered after this (Chap. 5) before the conclusion.

References

Agudo-Peregrina ÁF, Navío-Marco J (2016) Extended framework for the analysis of innovative smart city business models. In: 27th European regional conference of the international telecommunications society (ITS), Cambridge, UK, 7–9 Sept 2016. https://hdl.handle.net/10419/148654

Aletà NB, Alonso CM, Ruiz RMA (2017) Smart mobility and smart environment in the Spanish cities. Transp Res Proc 24:163–170. https://doi.org/10.1016/j.trpro.2017.05.084

Causone F, Sangelli A, Pagliano L, Carlucci S (2017) An exergy analysis for Milano smart city. Energy Procedia 111:867–876. https://doi.org/10.1016/j.egypro.2017.03.249

Galvão JR, Moreira L, Gaspar G, Vindeirinho S, Leitão S (2017) Energy system retrofit in a public services building. Manag Environ Qual 28(3):302–314. https://doi.org/10.1108/MEQ-02-2014-0028

March H, Ribera-Fumaz R (2016) Smart contradictions: the politics of making Barcelona a self-sufficient city. Eur Urban Reg Stud 23(4):816–830. https://doi.org/10.1177/0969776414554488

Mendoza J-MF, Sanyé-Mengual E, Angrill S, García-Lozano R, Feijoo G, Josa A, Gabarrell X, Rieradevall J (2015) Development of urban solar infrastructure to support low-carbon mobility. Energy Policy 85:102–114. https://doi.org/10.1016/j.enpol.2015.05.022

Pollio A (2016a) Technologies of austerity urbanism: the "smart city" agenda in Italy (2011–2013). Urban Geogr 37(4):514–534. https://doi.org/10.1080/02723638.2015.1118991

Pollio A (2016b) Smart cities as hacker cities. Organized urbanism and restructuring welfare in crisis-ridden Italy. Nóesis 25:31–44. https://dx.doi.org/10.20983/noesis.2016.12.3

Ramaswami A, Russell AG, Culligan PJ, Sharma KR, Kumar E (2016) Meta-principles for developing smart, sustainable, and healthy cities. Science 352(6288):940–943. https://doi.org/10.1126/science.aaf7160

Chapter 2
Low Carbon Cities

Abstract This chapter provides a background on the emergence of low carbon cities from urban planning, eco-city design, and green growth perspectives in the contemporary literature. Low carbon urbanism and eco-cities can be viewed as precursors to the development of smart cities, which—from an urban sustainability standpoint—have evolved through piecemeal automation and increasing integration as part of smart development. It is important to frame such developments from a social justice perspective to acknowledge that technology is working for humanity—to improve human quality of life and well-being.

Keywords Low carbon transitions · Smart development · Eco-cities · Green growth · Social justice · Energy poverty

There are strong implications for cities in the quest to curb carbon-based energies. As a concentration of 'activities, people, and wealth in limited areas' (Hallegatte et al. 2011), cities are both important generators of carbon dioxide or CO_2 emissions and end-users of goods and services, the production of which involves emissions elsewhere. Decreasing end-use energy demands through energy saving and efficiency measures alleviates the need to generate as much energy and, thus, moderates the carbon Footprint. But even with efficiency measures, some demands for energy will always be present while a growing population and economic development bring further pressures. It is necessary to decouple future economic growth from growing carbon emissions by decreasing the relative share of fossil fuels. Ultimately, carbon neutral or zero-carbon cities are one way that has emerged for expressing the net-zero carbon balance of cities in an effort to mitigate anthropogenic emissions of greenhouse gases or GHGs.

However, the transition to low carbon cities cannot be considered as smooth, linear, or even uncontroversial. The very magnitude of the task involved is the one that requires long-term and persistent political, economic, and institutional commitments as well as innovative, creative, and often 'alternative' ways of carrying on businesses producing and consuming goods and services. While some may argue that the community-level elements of a post-carbon society encourage a European social tradition that promotes individual freedom and social responsibility, human and social rights, balanced social and market models, and establishes cooperation and

M. J. Thornbush and O. Golubchikov, *Sustainable Urbanism in Digital Transitions*, SpringerBriefs in Geography, https://doi.org/10.1007/978-3-030-25947-1_2

peace (Carvalho et al. 2011), others are concerned that the new global 'consensus' about reducing global CO_2 is not sufficiently engaged with the principles of participatory democracy, while the political economy of low carbon transitions is itself not immune to sociopolitical struggles—the imposition of particular development strategies for cities, for example, often confronts the principles of social justice, equity, affordability, and civic participation, thus, undermining rather than reinforcing the more general principle of sustainability.

2.1 Cities, Energy, and Climate

'Climate' and 'energy' are two concepts that we encounter regularly in our life. However, it is the combined use of 'climate and energy' that has emerged as a policy-charged collocation to invoke the unity of the energy and climate change agendas. Indeed, the contribution of the energy sector to the world's anthropogenic GHG emissions is estimated to be three-fourths of the total emissions. This is mostly a result of the combustion of fossil fuels (coal, oil, gas, etc.), which is the main source of CO_2—the most prevailing GHG. The so-called decarbonisation of economies—i.e. reducing CO_2 emissions via limiting energy consumption and switching to non-carbon-based fuels—has become a major preoccupation of policies in the emerged consensus about the urgency of climate change, including at the urban level.

Meanwhile, urban communities are themselves vulnerable to the negative impacts of climate change. Urban areas concentrate people and infrastructure, often in hazard-prone areas. They experience some of the largest impacts from both gradual climatic changes and abrupt weather occurrences, and it is the poorer and socially deprived populations who usually suffer most. There is, therefore, a need for cities to embrace socially oriented policies of improving resilience and preparedness to cope with the negative environmental impacts.

The urgency of the climate and energy agenda for cities stimulated a surge in reports on cities and climate actions from major international organisations at the turn of the first decade of this century (UNEP SBCI 2009; OECD 2010a; World Bank 2010a; Bose 2010; UN-Habitat 2011). Many stress the need to address mitigation and adaptation efforts at the urban scale because of the potential to implement programmes effectively, concentrate people and industries, while providing new ideas and innovation that can spread quickly. Both sides of climate policy—mitigation (locally reducing the causes of the climate change) and adaptation (addressing the local negative impacts of climate change)—are considered to be the integral parts of a comprehensive urban strategy for climate neutrality (Golubchikov 2011). Such a strategy suggests that cities not only aim to achieve net-zero emissions of GHG by reducing such emissions as much as possible while offsetting the remaining unavoidable emissions, but also that cities aim to become future-proof, or resilient to the negative impacts of the changing climate, by improving their adaptive capacities.

However, Dodman (2009) argues that 'attempts to blame cities for climate change serve only to divert attention from the main drivers of GHG emissions—namely unsustainable consumption, especially in the world's more affluent countries'. He conveys on the basis of his research in the UK that GHG emissions are highest per capita in the rural northeast as well as Yorkshire and the Humber and not London and the West Midlands, which are both highly urbanised. Dodman (2009, p. 196) blames high consumption lifestyles in high-income countries—for example, the USA and Canada account for a fifth of global GHG emissions. America's 'throwaway economy' is considered to be a major contributor to climate change through waste production and the release of GHGs (Sheehan & Spiegelman 2010). For example, 44% of these emissions were found to come from the provision, use, and disposal of products and packaging, which is more than the energy used in buildings, transport, and in the provision of food (p 4). A large disparity in wealth even more remarkable within nations, as for example in India, where people with relatively high earnings generate four times more carbon per year than people who earn less per month (Dodman 2009, p. 197). This argument is also set forth by Satterthwaite (2008), who notes that individual and institutional consumption needs to be considered as drivers of generation (based on demand). For example, as industries are moving out of cities, this establishes a disjuncture between the spatial situation of demand and production, even though they may both be triggered by consumption in cities. Wood (2007) is similarly critical of urban development that continues to be inspired by the 'profligate lifestyle', which he argues was the main cause of the problem to begin.

Not all cities, however, follow identical pathways. Lankao (2007) espouses that carbon emissions per capita in cities is very small in low- and middle-income nations in comparison with wealth urban areas. She shows that per capita CO_2-equivalent emissions for American cities, such as Austin, Boulder, Santa Monica, and Berkeley, and some European cities, including Berlin, are greater than for Mexico City and Rio de Janeiro. This suggests that, for developing countries, including Latin American cities, a priority may be on coping with the implications of air pollution on human health and adaptation to the impacts of climate change, rather than curbing carbon emissions. Furthermore, cities in Latin America may not be able to adopt ecological modernisation as a suitable framework for addressing their environmental problems due to an emphasis on the industrial and technological change that overlooks the social and political context of an ecological switchover. The author further argues that even though eco-cities stress social and institutional dimensions of sustainability, they still reflect post-modern values that are best-suited to urban development in European and North American cities. Her final stipulation is that equity be considered as a carbon-relevant issue, since the wealthy in Latin American cities (and elsewhere, as relayed by others, e.g. Dodman 2009) have higher carbon emissions per capita and are more able to invest in their own well-being, as for example in paying for healthcare to remedy any impacts from poor air quality.

Even though lessons can be learned across countries, individual countries will achieve low carbon development (LCD) from different routes based on their national reality and development prospects as well as aspirations and capacities (Mulugetta & Urban 2010). For this reason, tailored approaches are necessary for low-, middle-, and high-income countries (World Bank 2010b, p. 204). Still, developing countries cannot undergo modernisation like developed nations. A case in point is China that has received attention as a nation that is at a critical period of industrialisation and urbanisation (Zhu & Shang 2010). Urbanisation can be taken as an opportunity for LCD. It is imperative that China develops a low carbon industrialisation model, since it must coordinate economic development and emissions control whilst continuing to industrialise and modernise (He et al. 2010).

2.2 Eco-Cities

The potential of urban planning is also realised in designing new low carbon or even zero-carbon cities or urban districts worldwide. Some prominent examples of low carbon communities—some completed, others only partly completed or uncompleted—have included Masdar City, still being built in Abu Dhabi as a zero-carbon, zero-waste, car-free municipality for 50,000 residents, intended to become the world's first climate neutral city; Dongtan in China remains an unrealised urban utopia, originally planned as a low carbon city to accommodate 0.5 million people; and the Western Harbour (Västra Hamnen) district of Malmö, turned from a brownfield site into an environmentally-friendly town based on 100% renewable energy. Smaller scale examples are BedZED—or Beddington Zero-Energy Development—consisting of 99 homes, which was the first zero-energy, low-impact, car-discouraging residential community in the UK; and similarly Etten-Leur, with 43 houses is a similar zero-energy housing demonstration project in the Netherlands. While these have served as encouraging examples, it is even more important to act in existing urban districts, where there is a large potential for paving a more sustainable future through climate smart urban planning. Indeed, most of those cities that embrace policies for carbon reduction make a proactive use of the instrument of planning.

Eco-towns and eco-cities are appealing in the context of urban sustainability as practitioners turn to paradigms or movements for a practical application of their ideas. For example, Roseland (1997) refers to activists among designers, practitioners who address sustainability, and visionaries, including bio-regionalists, social ecologists, and environmentalists at large, who are writing about green cities, eco-cities, and eco-communities in an Ecotopian view of social structure that is based on biophysical and social sustainability. From a planning point-of-view, Selman (1995) presents a case for sustainability planning that he postulates has been driven by 'the eventual acceptance, at the very highest levels of government, of the validity of environmentalists' claims about the potentially irreversible deterioration of natural capital stocks' (p 290). As a case-in-point, Knight (2010) writes about PlanIT Valley near Paredes in northern Portugal, where an eco-city with ambitions to be an environmentally sustainable city was being built. She relays that this development would occur before Masdar City. An advantage of PlanIT Valley over Masdar, for example, is its proximity to existing transport links. Importantly, it has a so-called brain that comprises a network of sensors operating like a nervous system to regulate water use and energy consumption in a kind of urban metabolism approach. Its buildings are designed in a hexagonal shape in order to most efficiently use space.

Joss (2010) provides profiles of eco-cities in a report of a global survey from 2009, where he maps, analyses, and compares 79 eco-city initiatives. His findings indicate that (at the time of his report) most eco-cities were situated in Europe (34 out of 79 in his analysis), particularly in places like the UK, Germany, and Scandinavian countries. But these projects appear in disparate countries, from China, Kenya, Japan, South Korea, and South Africa to Canada, Germany, Great Britain, Sweden, and the USA. In his opinion, the most original eco-city projects are situated in the Middle East and East Asia. According to Joss (2010, p. 245), these developments can be classified as: (1) new development (built from scratch); (2) expansion of existing urban area (e.g. new district or neighbourhood); and (3) retrofit development (within existing urban infrastructure). These initiatives have sprung up since the mid-2000s. More recently, Joss (2015) examines the smart city, which he advocates has rapidly been popularised since the late 2000s (as per our Chap. 1). Key criteria defining eco-cities include a substantial scale (in terms of area, infrastructure, and innovation); they normally occur across sectors (e.g. housing, transport, energy, waste, water, land, etc.); and are formulated as, embedded in, and supported by policy. Finally, Joss (2010) identifies various factors that are driving eco-cities, among them challenges from global climate change, rapid urbanisation, and socioeconomic regeneration. One critique of eco-towns in the UK is that they are really not entirely different from the new town programme (of which Milton Keynes is an example) that are based on utopian ideals and take a clean sweep approach to urban development (Smith 2009). It is hoped by government, nonetheless, that there will be a trickle-down effect of eco-towns that will influence urban development in towns and cities.

Retrofit development is the most effective way to pave a pathway towards zero-carbon cities. However, new developments, despite being criticised for being disintegrated, can also be effective if they extend their positive influence into the region. Reiche (2010) discusses a case study is the Masdar Initiative, which was developed by Masdar City and the Masdar Institute as a regional economic development programme announced in April 2006 by the Abu Dhabi government to promote sustainable energy. An ambition of this initiative is to boost renewable energy generation in Abu Dhabi. Even though it cannot be considered a pioneering project, since carbon neutral villages already exist, the size of the project is what sets it apart in the world. It has already influenced politics at various scales (domestic, regional, and global) and regional diffusion is suspected, with a plan for a sustainable campus at King Abdullah University of Science and Technology in Saudi Arabia and policy initiatives in Dubai. Also, its global media coverage has increased public awareness of renewable energy throughout the world. Masdar City has been developed to be self-sustaining, drawing its energy from renewable energy and waste-to-energy technologies, and any excess energy will feed the national grid (Crampsie 2008). Carbon capture and storage is one if its CO_2 emissions reduction projects, with storage in a national network for enhanced oil recovery (CCS-EOR), which could reduce the United Arab Emirates's annual CO_2 emissions by 40% and increase oil production by 10% as well as liberating natural gas now used to maintain oil-field pressure. The

CCS-EOR pipeline is expected to recover some 75 million tonnes of carbon per year from various sources across the country, connecting multiple industries. Therefore, Masdar will have an impact far beyond its borders.

Stockholm has a reputation of being innovative in integrating urban planning with sustainable waste, water, and energy management. However, Yeang (2010) states that '[e]cological design is still very much in its infancy. The totally green building or green eco-city does not yet exist. There is still much theoretical work, technical research and development, environmental studies and design interpretation that needs to be done and tested before we can say we have achieved a green built environment' (p 158). The author considers five design strategies to achieve stasis between the natural and built environments, the first of which focuses on green design that comprises eco-infrastructures, including nature's own utilities (green), engineering (grey), water management (blue), and the built environment (red). In addition, the author addresses bio-integration (of synthetic and natural environments), eco-mimesis (where design is inspired by ecosystems, like biomimicry), a restoration of impaired environments, and self-monitoring eco-designs.

2.3 Governing Low Carbon Transitions

Societal and behavioural changes can help achieve low carbon futures. Small towns, for example, deserve more attention than they have received so far, mainly due to the emphasis on large cities and city regions (Mayer & Knox 2010). Four movements have developed (local, organic and slow food, environmentalism, and entrepreneurship and creativity) that could help to bring out smaller places. Behavioural change is also expanded by Heiskanen et al. (2010) in the context of emerging low carbon communities. Even though such practical examples exist, researchers have unravelled that, at least for the London borough of Islington, pro-environmental attitudes and behaviour already existed in participants in a Green Living Centre that did not necessarily reflect upon other residents from the larger community, whose enthusiasm for sustainability change and interest in such community schemes were more mixed (Peters et al. 2010). It is suggested that this lack of interest could be counteracted by awareness-raising efforts that extend to the greater community. Findings such as these stress the need for social inclusion, particularly in community-based action to low carbon cities. Hence, it may not be as easy as identifying green niches at the policy level in order to encourage bottom-up (community action-based) governance (Seyfang 2010), when some members of the community do not participate. Perhaps one approach to encourage social participation in green initiatives could be achieved through the provision of rewards, such as funding to consumers, as of decentralised renewable energy systems (DRES) suggested by Williams (2010), who also argues that it is the producers of DRES that need to be targeted, rather than just consumers, and tightly regulated to install DRES, particularly in new housing developments.

But these issues of (the limits of) 'organic' governance bring us to considering the institutional structures governing change more generally. Despite the varied bottom-up movements, it is rather clear that systematic change lies with the institutional transformations driven by larger-scale actions and normative practices. Policies of decarbonisation are usually perceived to be in the hands of national and international institutions—although there has recently been a definite shift in international policy discourse from a dominance of top-down to multilevel governance, where the role of urban and regional institutions is often considered to be the key (Bulkeley 2009). Indeed, subnational and local governments can execute their authority in regulatory and planning functions, local charges, procurement procedures, and direct management of the public property. They not only translate national policy and resources into implementing policies 'on the ground', but also present themselves as an important vehicle for innovation in climate policy and practice.

An examination of implementing climate protection through urban planning for development and energy conservation in Newcastle-upon-Tyne and transport planning in Cambridgeshire found that sustainability is shaped by governance that extends across geographical scales and urban boundaries (Bulkeley & Betsill 2005). Cooperation between neighbouring municipalities is important because many initiatives cross the borders of individual administrative units (e.g. infrastructural projects or public transport). Here, the role of regional (subnational) administrations as coordinating, enabling, and funding bodies cannot be overstated (Wheeler 2009). Cities with a 'regional' administrative mandate, which is often the case for larger cities, are more capable of facilitating larger projects and territorial cohesion (OECD 2010b). It is not necessary, however, that city governments form only 'local' or 'regional' institutions. They can also create 'horizontal' national and international networks or associations that complement the 'vertical' regimes of governance. Such interurban associations provide a platform for sharing knowledge and for mutual support, and climate protection measures advocated by these associations are often expressed in agreements. An example is the 2007 World Mayors and Local Governments Climate Protection Agreement, which calls for a reduction in GHG emissions by 60% from 1990 levels worldwide by 2050 and, in industrialised countries nationally, by 80% from 1990 levels. It also declares a number of commitments for the signatories themselves, although without specific measurable targets.

Yet, long-term planning could be adopted towards a technology-explicit bottom-up approach (Bhatt et al. 2010) that is community-led (rather than government-led). The latter conveys a possible shifting in leadership from traditional routes. Others have discovered that grassroots action has the potential of building community capacity to accommodate low carbon practice that is place-specific (Middlemiss & Parrish 2010).

Apart from the cooperation between cities and between different administrations, city governments can seek a broader participation of stakeholders and the involvement of the population in climate-related decision-making processes in order to inform, and to be informed by, the local community's knowledge about climate challenges (including information about existing impacts on residents) and to share the ownership of new strategies with a larger group of stakeholders, thus, ensuring their more successful implementation.

Participation and cooperation can also help to bring in missing technical expertise. For example, universities represent an intellectual resource at the local level that can, on the one hand, support city governments in developing energy-related and carbon reduction policies and strategies and, on the other, play a key role in building knowledge on climate smart practices through changes in curriculum and teaching methods (World Bank 2010b). Indeed, a common approach in governance is establishing a science-policy competence, whereby decision-makers and experts come together to exchange information (Corfee-Morlot et al. 2011). For example, urban energy systems should be approached from sociotechnical perspectives that bring together experts in the sciences and social sciences, such as engineers (van der Sanden & van Dam 2010). Research and development efforts can recognise this contribution of multidisciplinary groups to be able to provide more diverse (holistic) solutions. Since technology is not a panacea, it is necessary to include linked scientific and social research that does not exclude the role of the individual, culture, and society. This is especially relevant since social participation and consumption can be driving forces of social change, which could affect the acceptance and adoption of new technology.

Speaking of local-scale actions, it seems that key factors for effective climate policy development and implementation in cities are collective public awareness and individual political leadership (Golubchikov 2011). Because the combination of these factors varies between different areas, there may be a large spectrum of responses among cities even within the same subnational jurisdiction or in smaller countries (e.g. for the case of Sweden, see Langlais 2009). Even those local governments that demonstrate proactive strategies often face a lack of legal mandate from national governments to implement advanced measures (OECD 2010b). This may include, for example, limited regulatory and fiscal authority, and lack of control over energy utilities or over strategic transportation development. In their strategies, local governments often go beyond their legislated capacity, which raises concerns over their effective implementation. Moreover, local responses to climate change are often circumscribed by the fiscal capacities of municipalities or regions. Even if substantial achievements can be reached with moderate cost, systematic and comprehensive climate policies are capital intensive. City governments need to identify sustainable sources of income for these policies. Local fiscal and payment regimes may themselves play a stimulating role to encourage or discourage certain activities, projects, or lifestyles, and these may have serious implications for climate neutrality. Some examples are public transport fees, parking fees, congestion charges, property taxes, and development charges. Financial resources can also be sought from the private sector; public-private partnerships may be established in order to share risk and raise private finance for infrastructure and energy efficiency projects. In their turn, national governments must ensure adequate resource mobilisation for local and regional governments, as it is at the national level that different forms of taxes can be institutionalised more comprehensively and effectively.

It is crucial that cities involve as many stakeholders as possible, including community and grassroots groups, academics, businesses, and activists, to promote a broader acceptance of policies and support for their implementation. The problem with eco-city approaches to LCD is that it may exclude low-income people, such as those living in urban slums. It is important for these citizens to gain the 'right to the city' through an avoidance of selective benefit-sharing, marginalisation, and discrimination that is evident in today's cities (UN-Habitat 2008). Cities in developing countries have a notable problem of a divide between the rich and poor, resulting in over 900 million people living in slums around the world (Antesberger et al. 2004). This is evident in megacities, such as São Paulo in Brazil. The poor occupying slums and squatter settlements in low-income countries often occupy risk-prone areas that are vacant and available to establish makeshift residences. These areas are commonly flood-prone, as for example in Indore, India and cities in Africa, such as Accra, Kampala, Lagos, Maputo, and Nairobi (Satterthwaite et al. 2007). The most effective upgrading programmes for slum and squatter settlements comprise local (government) support and community-based adaptation, which includes protection against extreme weather, and climate change by extension, as well as everyday hazards (Satterthwaite et al. 2009). Good planning and governance of towns and cities are important towards mitigation efforts and adaptation.

Many organic/grassroots movements advocate 'simple living' as a low-consumption approach to an urban lifestyle (e.g. Thornbush et al. 2013). This approach recognises that purchasing green (energy-efficient) products still contributes to the carbon emissions associated with the demand for production. Simple living entails buying what is necessary rather than in excess. This lifestyle, however, is very difficult to achieve, especially when consumerism is spreading rapidly in the world, causing hunger for energy production to fulfil several functions. The next chapter focuses on energy generation and use within the context of energy transitions.

References

Adger WN, Dessai S, Goulden M, Hulme M, Lorenzoni I, Nelson DR, Naess LO, Wolf J, Wreford A (2009) Are there social limits to adaptation to climate change? Clim Change 93(3–4):335–354. https://doi.org/10.1007/s10584-008-9520-z

Alberti M, Marzluff JM, Shulenberger E, Bradley G, Ryan C, Zumbrunnen C (2003) Integrating humans into ecology: opportunities and challenges for studying urban ecosystems. BioScience 53(12):1169–1179. https://doi.org/10.1641/0006-3568(2003)053%5b1169:IHIEOA%5d2.0.CO;2

Antesberger W, Arnoldt T, Bastiani A, Berz G, Domke A, Dürr M, Hackl S, Lahnstein C, Loster T, Oberhäuser J, Raduski B, Schmidt M, Schmieder J, Smolka A, Wulff R (2004) Megacities—Megarisks Trends and Challenges for Insurance and Risk Management. Münchener Rückversicherungs-Gesellschaft, München, p 79. https://www.preventionweb.net/files/646_10363.pdf

Bhatt V, Friley P, Lee J (2010) Integrated energy and environmental systems analysis methodology for achieving low carbon cities. J Renew Sustain Ener 2(031012):19. https://doi.org/10.1063/1.3456367

Boardman B (2010) Fixing fuel poverty: challenges and solutions. Earthscan, London. https://doi.org/10.4324/9781849774482

Bose RK, ed. (2010) Energy efficient cities: assessment tools and benchmarking practices. The International Bank for Reconstruction and Development/The World Bank,

Washington DC, p 227. http://documents.worldbank.org/curated/en/602471468337215697/pdf/
544330PUB0EPI01BOX0349415B01PUBLIC1.pdf
Bulkeley H (2009) Planning and governance of climate change. In: Davoudi S, Crawford J,
Mehmood A (eds) Planning for climate change: strategies for mitigation and adaptation for
spatial planners. Earthscan, London, p 284–296. https://doi.org/10.4324/9781849770156
Bulkeley H, Betsill MM (2005) Rethinking sustainable cities: multilevel governance and
the 'urban' politics of climate change. Environ Polit 14(1):42–63. https://doi.org/10.1080/
09644010420003l0178
Carvalho MG, Bonifacio M, Dechamps P (2011) Building a low carbon society. Energy
36(4):1842–1847. https://doi.org/10.1016/j.energy.2010.09.030
Corfee-Morlot J, Cochran I, Hallegatte S, Teasdale PJ (2011) Multilevel risk governance and urban
adaptation policy. Clim Change 104(1):169–197. https://doi.org/10.1007/s10584-010-9980-9
Crampsie S (2008) City of dreams. Eng Technol 3(15):50–55. https://doi.org/10.1049/et:20081509
Crichton D (2007) What can cities do to increase resilience? Philos T R Soc 365(1860):2731–2739.
https://doi.org/10.1098/rsta.2007.2081
Dodman D (2009) Blaming cities for climate change? An analysis for urban greenhouse gas emissions inventories. Environ Urban 21(1):185–201. https://doi.org/10.1177/0956247809103016
Dulal HB, Brodnig G, Onoriose CG (2011) Climate change mitigation in the transport sector through
urban planning: a review. Habitat Int 35(3):494–500. https://doi.org/10.1016/j.habitatint.2011.02.
001
Golubchikov O (2010) Action plan for energy-efficient housing in the UNECE region. United
Nations Economic Commission for Europe (UNECE), Geneva, p 57. https://www.unece.org/
fileadmin/DAM/hlm/documents/Publications/action.plan.eehousing.pdf
Golubchikov O (2011) Climate neutral cities: how to make cities less energy and carbon intensive
and more resilient to climatic challenges. United Nations Economic Commission for Europe
(UNECE), Geneva, p 85. https://www.unece.org/fileadmin/DAM/hlm/documents/Publications/
climate.neutral.cities_e.pdf
Hallegatte S, Henriet F, Corfee-Morlot J (2011) The economics of climate change impacts and
policy benefits at city scale: a conceptual framework. Clim Change 104(1):51–87. https://doi.
org/10.1007/s10584-010-9976-5
He J, Deng J, Su M (2010) CO_2 emission from China's energy sector and strategy for its control.
Energy 35(11):4494–4498. https://doi.org/10.1016/j.energy.2009.04.009
Heiskanen E, Johnson M, Robinson S, Vadovics E, Saastamoinen M (2010) Low carbon communi-
ties as a context for individual behavioural change. Energ Policy 38(12):7586–7595. https://doi.
org/10.1016/j.enpol.2009.07.002
Holling CS (1973) Resilience and stability of ecological systems. Annu Rev Ecol Sys 4(1):1–23.
https://doi.org/10.1146/annurev.es.04.110173.000245
Hunt A, Watkiss P (2011) Climate change impacts and adaptation in cities: a review of the literature.
Clim Change 104(1):13–49. https://doi.org/10.1007/s10584-010-9975-6
Joss S (2010) Eco-cities—a global survey 2009. WIT Trans Ecol Envir 129:239–250. https://doi.
org/10.2495/SC100211
Joss S (2015) Smart cities: from concept to practice. First published in the BSI (British Standards
Institution) Newsletter, February 2015. https://www.westminster.ac.uk/file/41956/download
Knight H (2010) The green city that has a brain. New Sci 2781:22–23
Langlais R (2009) A climate of planning: Swedish municipal responses to climate change. In:
Davoudi S, Crawford J, Mehmood A (eds), Planning for climate change: strategies for mitigation
and adaptation for spatial planners. Earthscan, London. https://doi.org/10.4324/9781849770156
Lankao PR (2007) Are we missing the point? Particularities of urbanization, sustainability and
carbon emissions in Latin American cities. Environ Urban 19(1):159–175. https://doi.org/10.
1177/0956247807076915
Mayer H, Knox P (2010) Small-town sustainability: prospects in the second modernity. Eur Plan
Stud 18(10):1545–1565. https://doi.org/10.1080/09654313.2010.504336

Middlemiss L, Parrish BD (2010) Building capacity for low-carbon communities: the role of grass-roots initiatives. Energ Policy 38(12):7559–7566. https://doi.org/10.1016/j.enpol.2009.07.003

Mulugetta Y, Urban F (2010) Deliberating on low carbon development. Energ Policy 38(12):7546–7549. https://doi.org/10.1016/j.enpol.2010.05.049

North P (2010) Unsustainable urbanism? Cities, climate change and resource depletion: a Liverpool case study. Geogr Compass 4(9):1377–1391. https://doi.org/10.1111/j.1749-8198.2010.00371.x

OECD (2010a) Cities and climate change. Organisation for Economic Co-operation and Development (OECD), Paris, p 19. https://www.oecd.org/env/cc/Cities-and-climate-change-2014-Policy-Perspectives-Final-web.pdf

OECD (2010b) Green cities programme. Organisation for Economic Co-operation and Development (OECD), Paris, p 4. https://www.oecd.org/regional/greening-cities-regions/46811501.pdf

Peters M, Fudge S, Sinclair P (2010) Mobilising community action towards a low carbon future: opportunities and challenges for local government in the UK. Energ Policy 38(12):7596–7603. https://doi.org/10.1016/j.enpol.2010.01.044

Pickett STA, Cadenasso ML, Grove JM (2004) Resilient cities: meaning, models, and metaphor for integrating the ecological, socio-economic, and planning realms. Landscape Urban Plan 69(4):369–384. https://doi.org/10.1016/j.landurbplan.2003.10.035

Reiche D (2010) Renewable energy policies in the Gulf countries: a case study of the carbon-neutral "Masdar City" in Abu Dhabi. Energ Policy 38(1):378–382. https://doi.org/10.1016/j.enpol.2009.09.028

Roseland M (1997) Dimensions of the eco-city. Cities 14(4):197–202. https://doi.org/10.1016/S0264-2751(97)00003-6

Satterthwaite D (2008) Cities' contribution to global warming: notes on the allocation of greenhouse gas emissions. Environ Urban 20:539–549. https://doi.org/10.1177/0956247808096127

Satterthwaite D, Huq S, Reid H, Pelling M, Lankao PR (2007) Adapting to climate change in urban areas: the possibilities and constraints in low- and middle-income nations. In: International institute for environment and development (IIED), London, p 112. https://pubs.iied.org/pdfs/10549IIED.pdf

Satterthwaite D, Dodman D, Bicknell J (2009) Conclusions: local development and adaptation. In: Bicknell J, Dodman D, Satterthwaite D (eds) Adapting cities to climate change: understanding and addressing the development challenges. Earthscan, London, pp 359–383. https://doi.org/10.1080/17535069.2013.846001

Selman P (1995) Local sustainability: can the planning system help get us from here to there? Town Plann Rev 66(3):287–302. https://doi.org/10.3828/tpr.66.3.j451854764541u7x

Seyfang G (2010) Community action for sustainable housing: building a low carbon future. Energ Policy 38(12):7624–7633. https://doi.org/10.1016/j.enpol.2009.10.027

Sheehan B, Spiegelman H (2010) Climate change, peak oil, and the end of waste. In: Heinberg R, Lerch D (eds) The post carbon reader: managing the twenty-first century's sustainability crises. Watershed Media, Healdsburg, CA, pp 363–384

Smith PF (2009) Building for a changing climate: the challenge for construction. Earthscan, London, Planning and Energy, p 200

Thornbush M, Golubchikov O, Bouzarovski S (2013) Sustainable cities targeted by combined mitigation–adaptation efforts for future-proofing. Sustain Cities Soc 9:1–9. https://doi.org/10.1016/j.scs.2013.01.003

UN-Habitat (2008) State of the world's cities 2010/2011: bridging the urban divide. United Nations Human Settlements Programme (UN-Habitat), Earthscan, Washington DC, p 244. https://unhabitat.org/books/state-of-the-worlds-cities-20102011-cities-for-all-bridging-the-urban-divide/#

UN-Habitat (2011) Cities and climate change: policy directions. United Nations Human Settlements Programme (UN-Habitat), Earthscan, Washington DC, p 279. https://unhabitat.org/books/cities-and-climate-change-global-report-on-human-settlements-2011/#

UNEP SBCI (2009) Buildings and climate change: summary for decision-makers. United Nations Environmental Programme (UNEP), Sustainable Buildings and Climate Initiative,

Paris, p 56. https://www.greeningtheblue.org/sites/default/files/Buildings%20and%20climate%20change_0.pdf

Van der Sanden MCA, van Dam KH (2010) Towards an ontology of consumer acceptance in socio-technical energy systems. IEEE Xplore (06 January 2011), In: Third international conference on infrastructure systems and services (INFRA): next generation infrastructure for eco-cities, 11–13 Nov 2010, Shenzhen, China, p 1–6. https://doi.org/10.1109/INFRA.2010.5679223

Wheeler S (2009) Regions, megaregions, and sustainability. Reg Stud 43(6):863–876. https://doi.org/10.1080/00343400701861344

Williams J (2010) The deployment of decentralised energy systems as part of the housing growth programme in the UK. Energ Policy 38(12):7604–7613. https://doi.org/10.1016/j.enpol.2009.08.039

Wood J (2007) Synergy city; planning for a high density, super-symbiotic society. Landscape Urban Plan 83(1):77–83. https://doi.org/10.1016/j.landurbplan.2007.05.006

World Bank (2010a) Cities and climate change: an urgent agenda. The International Bank for Reconstruction and Development/The World Bank, Washington DC, p 81. https://siteresources.worldbank.org/INTUWM/Resources/340232-1205330656272/CitiesandClimateChange.pdf

World Bank (2010b) Development and climate change. The International Bank for Reconstruction and Development/the World Bank, Washington DC, p 417. https://siteresources.worldbank.org/INTWDR2010/Resources/5287678-1226014527953/WDR10-Full-Text.pdf

Yeang K (2010) Briefing: strategies for designing a green built environment. Urban Des Plann 163(4):153–158. https://doi.org/10.1680/udap.2010.163.4.153

Zhu S, Shang T (2010) Low carbon urbanization way forward for two-oriented society. In: Proceedings of 2010 IEEE international conference on advanced management science, vol, 3, pp 246–249. https://doi.org/10.1109/ICAMS.2010.5553245

Chapter 3
Energy-Based Transitions

Abstract The focus of this chapter is on reducing energy consumption in cities, including through decarbonisation efforts and transitions as well as improved energy efficiency. Continued investments in the production of renewable energy sources and the transmission (or distribution) of green energy are needed in order to sustain a low carbon supply as for instance in district heating and cooling (DHC) and as combined heat and power (CHP) or cogeneration. In addition to investing in such technologies, it is also pertinent to reduce energy consumption and promote energy conservation. City challenges regarding emissions are addressed. It helps to have building standards in place and building code guiding sustainable homes as well as effective planning to direct development, even amid rapid development. Spatial planning has potential when deployed alongside building control. Decisions regarding density building, suburbs, and transport are vital to examine in the context of the New Urban Design as well as sustainable development.

Keywords District heating and cooling (DHC) · Combined heat and power (CHP) · Building code · Spatial planning · New urban design · New technologies · Sustainable development

Climate mitigation efforts have focused on the energy sector and transport sectors, built environment and densification, and urban greenery. This reflects the efforts in relation to both the supply and demand side of power. Specific attempts have been made in the energy sector, on the supply side, to improve energy generation efficiency, shifting to less carbon-intensive fuels, keeping electricity affordable, as well as developing public and public partnerships. Hydroelectricity, wind, solar photovoltaic, solar thermal, geothermal, tide, and wave are all renewable types of energy that do not involve direct greenhouse gas (GHG) emissions (albeit there are indirect emissions from building power installations). Biomass (wood, biofuels, waste) can also be a carbon neutral source of energy if the burned biomass is renewed in a sustainable way.

Many countries have made decarbonisation efforts, by increasingly using a greater share of fuels with a reduced carbon content and technologies with fewer emissions and generating a larger proportion of electricity and heat from non-fossil fuels. According to IEA (2018) data, in Iceland for example, 83% of total primary energy

M. J. Thornbush and O. Golubchikov, *Sustainable Urbanism in Digital Transitions*, SpringerBriefs in Geography, https://doi.org/10.1007/978-3-030-25947-1_3

supply (TPES) is derived from hydropower and geothermal power. Combined, these two types of power provide for all of the country's electricity needs (respectively, 75 and 25% of electricity generation). Geothermal power was found, in addition, to be responsible for 94% of the country's heat production. Under previous administrations, the USA formulated an extensive plan of decarbonisation that was based on proven technology and distribution systems (e.g. Shinnar & Citro 2006). The plan envisaged that in the next few decades, the USA would be switching to non-fossil energy sources, including concentrated solar thermal (CST), nuclear, geothermal and hydroelectric, wind, solar cells, and biomass energy. The plan did not include a major programme of carbon sequestration because it was considered to be more costly than CST and nuclear. Perhaps electricity is a suitable approach for infrastructure in a world currently in transition because it does not rely on the type of energy source, whether it is renewable or not, and is already broadly in use. This is advocated by Zerocarbonbritain2030 (ZCB2030) that proposes a future scenario, where '[t]he roads and rails will buzz with the sound of power lines, batteries and fuel cells' (Kemp & Wexler 2010, p. 105).

3.1 Urban Energy Infrastructure

Globally, there is not only an increased interest in renewable energy sources, but also in decentralised energy generation and distribution (Goodier & Rydin 2010). The call for low carbon energy offers opportunities to shift from 'large' vertically integrated energy industries to decentralised neighbourhood-scale generation, which can be sufficient to cover all local needs. Increasing use of decentralised energy is also a way to reduce energy transmission losses, since energy systems can be more efficient when power lines to consumers are as direct as possible and the number of transformation steps minimised. It is of course the city and regional levels that can play a key role in decentralised energy. Even when the city government does not own and operate power-generating facilities (although the opposite is often true), it can use a number of levers to promote local green energy infrastructure. For example, the city can purchase renewable energy for city operations; identify strategic sites where renewable and low carbon energy sources could be located; provide planning incentives and development land; permit the construction of only efficient and clean power installations; and require new developments to connect to district heating systems. In short, the following options are implemented at the city level for city-scale decentralised renewable and low carbon power supply (Golubchikov 2011):

(a) Switching to lower-carbon technologies and promoting district heating and cooling systems with cogeneration and tri-generation;
(b) Installing renewable power installations, e.g. wind turbines, solar farms, energy from biomass and waste plants;
(c) Promoting onsite microgeneration of heat and electricity in the buildings sector;
(d) Developing a smart grid and efficient municipal energy services.

The fuel mix used in power generation also matters. Increasing the share of gas in energy supply has been promoted in many cities; indeed, natural gas contains 40–50% less carbon than coal and 25–30% less carbon content than oil, only marginal quantities of sulphur, and is more energy-rich and efficient. Power stations with modern gas turbines can achieve 60–65% of conversion efficiency, but the most modern city-based gas-fired combined heat and power (CHP) plant can reach efficiencies of more than 90% at the point of end-use (due to lower losses from transmission, fewer condensation losses in boilers, and the close proximity to the consumers). It is evident that a considerable amount of primary energy and carbon emissions can be saved by the large-scale deployment of modern CHP plants. The CHP technology, which is also known as cogeneration, can be used for both industrial and non-industrial purposes and also at the micro (household) scale, but it is most advantageous if connected to district heating (also known as community heating) and deployed at a city- or neighbourhood-scale. In addition to satisfying local needs in heat, hot water, and power, CHP plants can provide cooling, by chilled water (this is known as tri-generation or as combined cooling, heat, and power).

Although district heating and CHP can function independently of each other, district heating and cooling (DHC) with CHP is today one of the most proven, efficient, and cheapest available technologies to reduce emissions and save energy at the city level. District heating, in particular, is considered for deployment in areas of high population densities with continuous demand. However, there are examples of countries where even low-density areas are supplied by district heating. Countries with significant shares of low-density, single-family houses connected to district heating in 2003 were Iceland (85%), Denmark (48%), Finland (13%), and Sweden (10%) (Nilsson et al. 2008). Remarkably, however, there is a strong opposition at the community level against district heating in countries such as the UK, which lack an appropriate tradition.

District heating and cooling can be designed as a flexible system, so that apart from CHP, DHC networks can be supplied from a variety of other sources, including geothermal and solar heating stations; fuel cells; biomass; surplus heat from industries; and energy from waste facilities. The ability to integrate diverse energy sources may provide for a flexible platform to reduce dependency on a single source of supply and to introduce competition into the supply chain. Similarly, CHP plants themselves can work on different fuel mixes. A challenge for climate neutral policies is to drive the whole energy infrastructure of district heating and CHP towards renewable supply; the anticipation of such a move should be integrated into the planning for new installations. For example, such 'future-proofing' has been a priority for London authorities planning for easy replacement or refuelling of new-build gas-powered CHP with renewable fuel or hydrogen in the future (Jones 2009).

Apart from cogeneration, cities promote other forms of renewable energy supply, such as city-scale or neighbourhood-scale power installations and even smaller (building-scale) microgeneration. Again, different sources of renewable energy are used—geothermal; wind; solar; ocean; biomass; landfill gas; and waste-to-energy. The small power generators can be linked to the common electricity grid and district heating or, alternatively, supply electricity and heat directly to the consumer (such as stand-alone renewable power operating at distribution voltage level). The introduction of electricity buy-back may promote renewable technologies in China, for instance, under a distributed energy system employing CHP, biomass energy, and photovoltaic technology (Ren et al. 2010). The city of Guelph (Ontario, Canada) has included wind energy in its community energy plans to produce energy within its municipal city boundaries. This scheme could generate between 8 and 29% of its total electricity demand from a baseline in 2005 (McIntyre et al. 2011, p. 1445). These trends will certainly affect the way cities are designed and planned. Microgeneration or onsite renewable energy generation in the buildings sector—both by commercial buildings and dwellings—is also increasingly promoted. Networked microgeneration might even be sufficient to cover all local electricity and heat demand, given that the final energy consumption is reduced through improving end-use efficiency. Microgeneration can include different types of heat pumps; small CHP plants; solar PV and thermal collectors; wood pellet stoves; small wind turbines; and other renewable technologies. For example, as part of the Hamburg Climate Action Policy for 2007–2012, Hamburg is carrying out a number of measures for the deployment of solar roofs. To this end, about 150,000 roofs in Hamburg were examined to determine their potential for generating energy, including by using a laser scanner flight programme, which measures both the direct and diffuse solar radiation potential of roofs in the city (City of Hamburg 2011).

In short, energy policy could provide a means by which to target energy consumption and the reduction of greenhouse gases or GHGs, as exemplified by Barcelona and London that are in carbon lock-in and using urban planning policies towards low carbon and renewable energy technologies in retrofitted and new buildings (Maassen 2010). One example is London's use of onsite decentralised energy generation in new developments. The project 'Challenging lock-in through urban energy systems (Clues)', for instance, raised important questions relating to urban areas in the UK and barriers to their energy systems, system decentralisation, aggregate decarbonisation targets, and sustainability (Rydin et al. 2010).

3.2   The Built Environment

Since cities are typically seen to be the largest source of carbon emissions (Hunt et al. 2007), they need to be redesigned to reduce their emissions. This can be achieved through strong legislation, especially building regulations, waste and water management as well as city planning, planned infrastructure changes, using the latest building technology and alternative (renewable) energy, such as solar and wind integrated into building design and retrofitted, local (in-city) power generation to reduce loss of energy via transportation, city-specific plans particularly for developing countries, and possibly the eventual abandonment of unsustainable (or highly vulnerable) cities.

There is no doubt that the built environment in particular poses a major challenge, especially since it is responsible for emitting over a third of the world's emissions and consumes about 40% of energy for residential and commercial buildings in Western societies alone (James 2009, p. 52), releasing much carbon dioxide or CO_2 into the atmosphere. Ürge-Vorsatz et al. (2007) postulate that it is possible to reduce these emissions by 30% (for a selection of best practices), having examined 60 policy evaluation reports, representing at least 30 different countries across four different continents. They discovered the most effective policy instruments in this sector to be appliance standards, building codes, tax exemptions, and voluntary labelling, which are found to be even more effective than Kyoto Protocol flexible mechanisms and carbon taxation. In their overall assessment, the most cost-effective instruments (of energy savings achieved with negative costs for society) are appliance standards, demand-side management programmes, and mandatory labelling.

A report by PRP Architects et al. (2008) examines six exemplary places with large-scale development situated from the city centre to suburbia. Namely Adamstown (near Dublin, Ireland), Amersfoort (the Netherlands), Freiburg (Germany), HafenCity (Hamburg, Germany), Kronsberg (Hanover, Germany), and Hammarby Sjöstad (Stockholm, Sweden) were subject. In Freiburg, people are using their cars less and either taking public transports or cycling. In terms of climate-proofing, there was a 60% reduction in CO_2 emissions in Kronsberg (p 18).

It is particularly beneficial for China's rapid urban development that considerable carbon emissions reduction can be gained from an approach that combines building design and construction along with urban planning and building material industries (Li et al. 2009). Even though China has adopted a low carbon economy (Jing et al. 2010), which is targeted to deploy clean energy, including renewable energy mainly derived from small and large hydro power plants (Zhang et al. 2010), it experiences barriers to carbon reduction, as in larger commercial buildings situated in Beijing and Shanghai that cannot employ energy managers of a sufficient calibre (Jiang & Tovey 2010). A study conveys that even though there was progress in carbon reduction in Chinese cities in the 1990s, this progress has slowed down or even reversed in recent years (Dhakal 2009). Urban management in China is restricted by departmentalisation as well as poor information sharing and coordination, creating an information isolation island problem (Wang & Cao 2010). Instead of investing in low carbon technologies, such as deriving renewable energy from wind or carbon sequestration, it was found that a better mitigation tool in terms of cost-effectiveness for China could be the improvements in building energy efficiency, like in cities located in northern China, such as Tianjin, which implemented building energy efficiency policies in its residential sector in the 1980s (Li et al. 2009). It has been postulated that new urban construction in China should move towards low carbon eco-city status (Li et al. 2010).

Other measures are suggested for existing buildings, for example Kelly (2009) advocates that: building fabric could be reengineered; appliance efficiency could be improved; electricity to homes could be decarbonised either through the grid or the use of renewable sources of energy; and changes in personal behaviour could provide solutions. Even historic buildings can be retrofitted, as with the case of The Hague's technological configuration (Peltier 2009), which did not require any major changes to its façade. Some European museums have adopted energy conservation techniques in their buildings in order to reduce energy consumption, including considerations of indoor temperature (heating and cooling) as well as lighting and other electrical devices (Zannis et al. 2006). It is possible to retrofit these non-domestic buildings, as also shown for Bristol, UK in attempts to reduce the (financial) risks associated with unsustainable buildings (Femenías & Fudge 2010). One of the most cost-effective solutions is insulation retrofitting, particularly with the use of nano insulation materials (Jelle et al. 2010).

A report by the Environmental Change Institute (Boardman et al. 2005), namely 40% House, was challenged by Power (2010), who opposes its proposal to achieve reductions in CO_2 emissions in building through the demolition of leaky homes. She advocates that the proposal is based on unsupported assumptions, including that new homes will have a better energy performance, and ignores both embodied energy and waste generated in the new building as well as the energy costs of infrastructure. An approach that is more conserving and recycles materials is likely to be more sustainable. For example, she suggests that 'higher refurbishment standards for existing homes using known methods (including under-floor and solid wall insulation) offer better value and potentially greater gains more quickly and cheaply than demolition and replacement buildings' (p 214). Research by Thornbush and Viles (2007) supports that old building stone, for example, is more stable and affected less (in terms of material loss) than newly exposed surfaces. This could indicate that there is some advantage to salvaging stone used in older constructions. Moreover, it is possible to improve existing building envelopes through measures taken to doors and entrances, draught-proof, window films and glazing, natural ventilation, solar shading, solar reflective surfaces, insulation, and green roofs (Rawlings 2010) as well as green walls. Using an optimum insulation thickness, for example, was found to reduce CO_2 emissions by 27% in Erzurum, one of Turkey's coldest cities (Çomakli & Yüksel 2004, p. 939).

Leaky homes can be improved with increased insulation, as also offered by vegetation. Green roofs integrate the positive effects of vegetation cover directly into the buildings' design. They reduce the over-heating of buildings in summer and provide a better thermal insulation in winter, thus, improving the building's own energy performance in addition to the positive effects for the neighbourhood as a whole. For example, traditional rooftops in North America and Central Europe can reach temperatures as high as 90°C during the summer, but green roof temperatures stay below 50°C. This demonstrates that the difference in surface temperature between a green roof and an unplanted roof can reach 40°C and more (Gartland 2008). A cooling roof is also beneficial for solar panels, as they currently work best at temperatures up to 25°C and have a reduced productivity at higher temperatures. Furthermore, green roofs intercept stormwater runoff and reduce the load on the building's drainage system, thereby extending its maintenance cycle. There are interesting examples of compulsory green roofs as posited by a recent by-law that requires the construction of green roofs on public and private buildings in the City of Toronto (Ontario, Canada). In Chicago (USA), government buildings require green roofs and cities in Austria, Switzerland, and Germany, following the original experiences of Basel and Linz, have introduced either compulsory requirements for greening all flat roofs on new buildings or additional subsidies for such measures for existing roofs (Golubchikov 2011).

This approach has the added benefit of urban greening, which will help to absorb atmospheric CO_2 through carbon capture by green façades and roofs in addition to other green spaces. Forests, for example, have been promoted as carbon sinks for low carbon cities (Jiang et al. 2010). They are believed to be an important strategy in global warming mitigation, moving towards the reduction in emissions associated with deforestation and degradation and the improvement of forest management and afforestation. This approach of urban greening has been recently extended to include urban agriculture (e.g., Thornbush 2015), such as food projects (Hopkins 2010). Living walls adopted in office buildings could also improve the quality of indoor circulated air and, hence, human health. Besides capturing CO_2 gas, plants are also capable of trapping particulate matter, which could reduce the incidence of human cancer in cities due to the inhalation of black carbon particulate.

3.3 Spatial Planning, Urban Density, and Mobility

Today, spatial planning in its various manifestations—regional and urban planning, land use zoning—finds itself right at the heart of adaptation and mitigation measures. Indeed, urban layout, public transit provision, and integrated district heat-electricity systems are some of the planning considerations that have long been acknowledged among the principal instruments to reduce urban energy intensity (e.g. Owens 1986). Planning is also instrumental in identifying risk-prone zones and providing spatial strategies to safeguard urban infrastructure. What is no less important is that planning decisions on land use and urban layout have impacts lasting for decades and even centuries. Particular land use and infrastructural patterns create the circles of 'path dependence', when future investments are predetermined by existing infrastructure, in this case, which may lock economies into particular lifestyles and patterns. Spatial planning is important to prevent being locked into high-carbon or hazard-prone conditions that would be expensive or impossible to alter later (World Bank 2008).

Spatial planning is relevant to all sectors of the urban economy and is principal for the integration of different sectors and urban systems into a consolidated spatial strategy (Rydin 2010). It is often the case, however, that links between territorial plans and climate policies are weak. This is because climate policies are often focused on particular economic sectors and may disregard spatial relations between and within urban sectors as well as the importance of how urban space is organised (OECD 2010). A purposeful integration of planning with policies for climate-smart growth is currently promoted in the context of climate change strategies.

Building control is a powerful tool to complement planning. Contrary to spatial planning itself, which may be opposed by some political ideologies as 'excessive' public interference (and, therefore, being limited in certain regions), building control is more easily accepted as a regulatory regime (this has been the case for the USA and some post-socialist countries; see Golubchikov 2004; Stanilov 2007).

Building control may also ensure the presence of planning targets in actual construction practice, including in the private sector. Legal provisions can be established such as those, for example, which require that building permits are only issued for projects that are optimised spatially to reduce energy demand, including density and transport considerations; taking advantage of natural heating, cooling, lighting, and shading potentials; and that incorporate building materials and other means for reducing urban heat island effects (e.g. cool walls, roofs and paving, increasing green areas). Moreover, urban development projects should be subject to a holistic assessment with regard to their environmental standards, which means that the full lifecycles of buildings (all stages from the manufacturing of construction materials to demolition and recycling of materials) are optimised in order to reduce the overall carbon Footprint.

Studies have found that multi-model land use and transportation design in planning for building improvements can reduce emissions; higher-density building is also important; energy efficiency can be achieved across a variety of building types; and affordable housing near work should reduce commuting costs (Condon et al. 2009). Research performed in the City of Toronto (Ontario, Canada) has broadly shown that urban form and density are important considerations (Norman et al. 2006). Policies that reduce operational energy and high-density development nearer to places of employment as well as increase the use of public transport and reduce private vehicle use in the suburbs should be given priority. Alternative fuels and renewable energy should be adopted in order to reduce transportation and operational energy use and GHG emissions from residential development. A study for the Chicago (USA) metropolitan area (Lindsey et al. 2011) has found that vehicle miles of travel, energy consumption, and CO_2 emissions from privately-owned vehicles are augmented with distance from the central business district, but reduced with residential density. This research suggests that high-efficiency vehicles may help to reduce emissions in cases of urban sprawl.

A modern approach to urban planning is the so-called New Urbanist design, which provides an alternative to conventional low-density development (Stevens et al. 2010). Steemers (2003), for instance, sees the benefits of a compact design for cities and towns with integrated public transport. Increased density is a part of this approach, which could use green standards at a lower cost (HTA et al. 2007). Some urban systems depend on achieving a critical density based on the mass of dwellings, such as the effective deployment of combined cooling heating power (CCHP) systems. Moreover, a sufficient volume of development would allow energy companies to support low carbon energy technologies that employ renewable sources of energy (wind, solar, woodchip, etc.). This combined with an integrated energy strategy, which includes a green transport plan, would go a long way to promote a low carbon lifestyle. For example, Power (2010, p. 206) specifies a home density of at least 50 homes per hectare, comprising some 110 people, over the current planning standard of 30 homes per hectare in order to maintain public transport (a regular bus service) as well as shops and schools in towns.

Many authors have also advocated such an approach towards sustainable development, where urban growth that is balanced, compact, and coordinated is geared towards achieving economic, social, and environmental benefits (Nadin 2006); as well as being aligned with the planning and decision-making process involving sociocultural, juridical, aesthetical, and ethical aspects (Vandevyvere & Stremke 2012). This can be attained through a more polycentric pattern in cities and towns and the prevention of urban sprawl. Urban planning needs to consider the size of the city and any associated characteristics of its residents. At a certain level of density, the negative environmental, energy, climate, and sociophysiological impacts start outweighing the gains. Super density also amplifies the negative effects of climate on cities—especially in areas with a high concentration of tall buildings (Roaf et al. 2009). Larger cities normally have larger surrounding areas and involve more long-distance travel, so that people's travel performance is connected to a country's spatial-economic organisation (Perrels 2008). For example, in Finland, medium-sized cities (of around 100,000 inhabitants) have the strongest mitigating effect on transport performance.

There is no consensus on what the optimal level of urban density actually is, nor on whether higher densities should always be encouraged. Moreover, key problems for intensified densities and the 'compact city' are that many cities already differ from an 'optimal' density and that the habits and aspirations of a considerable portion of the population are based on low-density models. There is, however, a broader consensus about the harmful effect of sprawl and the benefits of mixed-use development. The latter generally includes integrating housing, work, facilities, and entertainment in close proximity so that both trip distances and car dependence are reduced. Mixed-use development may also be accomplished in lower-density townscapes, so that existing low-density areas can be transformed towards mixed-use development, based on a strategy of stimulating urban polycentricity.

One case-in-point is the suburb. It is argued that measures associated with a compact city design have not been explicitly geared towards suburban areas, which have their own unique challenges in this transformation towards low or zero-carbon cities, including a slow pace of change (Williams et al. 2010). The built environment in the suburbs is also challenged by other problems revolving around the retrofitting of existing houses and fragmented property ownership and management. This was addressed by Rice (2010), who examines retrofitting existing suburbs towards sustainable urbanism using a compact city strategy that is promoted by the government of the UK. His analysis reveals that it is both feasible to retrofit the suburbs and that this endeavour is locally viable. It can even, in some cases, encourage more sustainable lifestyles, amongst them improved accessibility as well as social inclusion and even physical and mental health benefits.

There is also a broad consensus in the literature that public transport is a crucial consideration to curb emissions from travel. Developing countries, including China and India, are investing in public transport, such as city bus fleets that use alternative fuel in China (Ou et al. 2010). Indian cities such as Mumbai, with a higher share of public bus transport and suburban rail, has experienced a 60% reduction in energy and emissions compared to other cities like Delhi (Das & Parikh 2004), where future emissions (by 2020) are expected to be controlled by the adoption of efficient vehicles and fuels. The use of public transport (mass transit systems), rather than private passenger vehicles, can lead to energy savings in the transport sector for Bangkok due to less of an energy demand as well as reduced local air pollution, including carbon emissions (Phdungsilp 2010).

In addition to developing public transport and non-motorised transport infrastructure, transportation demand management includes optimising traffic flows. Improving the state of the road infrastructure and providing intelligent transportation system (i.e. using various forms of information and communication technologies for real-time information exchange between vehicles and road infrastructure) can reduce traffic bottlenecks and divert traffic from inner city areas. In this way, it helps to alleviate congestion and attendant air pollution, additional GHG emissions, and time losses. Speed limits can also be used, as high speeds lead to higher fuel combustion and, hence, amplified CO_2 emissions. Important options, which encourage modal shifts and rationalise transport flows, also include road pricing and car parking policies; congestion pricing tolls; park-and-ride facilities; ridesharing and car clubs; and travel planning. The promotion of remote forms of doing business and acquiring services (such as IT-based) in order to alleviate dependencies on traffic loads is also an important strategy.

Transport needs to undergo considerable change to accommodate the increasing number of city-dwellers and reduce reliance on private vehicles. There is an interesting trend, for example, of adopting aerial ropeways for urban transport. Many cities have such urban endeavours underway, including metro cables in Medellin and Caracas, Algeria's aerial ropeway serving the cities of Skikda and Tlecern, which is linked to their transit systems, and the new gondola system in Koblenz (UN-Habitat 2010). These aerial ropeways use less material and energy and are non-polluting. They have a small Ecological Footprint and are among the world's safest and most sustainable modes of transport. A relatively recent initiative by Google is to develop technology for self-driving vehicles (automated cars) that rely on video cameras, radar sensors, and lasers along with roadmaps to navigate through traffic. This could stimulate improved navigation, such as the shortest possible route taken in a single road trip as well as reduce road accidents.

Hydrogen is suggested as a sustainable transport fuel. For instance, Hart (2003) discusses a shift from transport energy derived from the burning of fossil fuels to hydrogen that is produced from renewable resources. According to him, this would reduce GHG emissions to zero and improve air quality, and even diminish noise pollution associated with the internal combustion engine. However, hydrogen energy infrastructure—the hydrogen road—implemented in Norway between Oslo and Stavanger, was affected by problems stemming from user technology, whereby sociotechnical networks failed usually due to technological immaturity (Kårstein 2010).

A need for a sociotechnical understanding of domestic consumption behaviour (particularly of the systems, standards, and norms that shape consumption) has been reinforced by others like Moloney et al. (2010), who analysed local carbon neutral community programmes in Australia. There is a business model for increasing the presence of electric vehicles (EVs) in private transport through an Electric Recharge Grid Operator that comes in advance of EVs in an intelligent rechargeable network that is based on renewable energy (Andersen et al. 2009). Some countries have already introduced this model, including Israel, Denmark, Australia, and the USA; it is in-line with a long-term goal for the automotive industry of zero emissions and a nil reliance on hydrocarbons for fuel (Sveum et al. 2007).

New technologies, however, are not a panacea. Satterthwaite (2011) argues that high standards of living can be achieved in cities with low GHG emissions through reduced resource use and waste, including lower material standards for the wealthy. In other words, adopting new (more energy-efficient) technologies is a consumerist approach to the problem, which could also be remedied by an alternative approach of reduced consumption. This would call for behavioural change, which could be reached at a lower cost. Such a non-consumerist approach could be useful, particularly for developing nations that cannot afford new technology and infrastructure. From the viewpoint of transport, changing people's travel habits from being convenience-oriented to low carbon-oriented could improve energy conservation based on behavioural change (Zhao & Chu 2009). Behavioural change can be established through policy, as through the provision of sustainable transport in order to reduce dependence on petroleum (Chapman 2007).

References

Andersen PH, Mathews JA, Rask M (2009) Integrating private transport into renewable energy policy: the strategy of creating intelligent recharging grids for electric vehicles. Energy Policy 37(7):2481–2486. https://doi.org/10.1016/j.enpol.2009.03.032

Boardman B, Darby S, Killip G, Hinnells M, Jardine CN, Palmer J, Sinden G (2005) 40% house. Environmental Change Institute, Oxford, p 126. https://www.eci.ox.ac.uk/research/energy/downloads/40house/40house.pdf

Chapman L (2007) Transport and climate change: a review. J Transp Geogr 15(5):354–367. https://doi.org/10.1016/j.jtrangeo.2006.11.008

City of Hamburg (2011) Hamburg climate action plan 2007–2012: update 2010/2011 (document 19/8311). Centre for Climate Issues, State Ministry of Urban Development and Environment, Free and Hanseatic City of Hamburg, Stadthausbrücke, p 40 + appendices. https://www.hamburg.de/contentblob/2982846/data/hamburg-climate-action-plan-2010-2011-english-version).pdf

Çomakli K, Yüksel B (2004) Environmental impact of thermal insulation thickness in buildings. Appl Therm Eng 24(5–6):933–940. https://doi.org/10.1016/j.applthermaleng.2003.10.020

Condon PM, Cavens D, Miller N (2009) Urban planning tools for climate change mitigation. Lincoln Institute of Land Policy, Cambridge, MA, p 48. https://www.lincolninst.edu/sites/default/files/pubfiles/urban-planning-tools-climate-change-mitigation-full_0.pdf

Dhakal S (2009) Urban energy use and carbon emissions from cities in China and policy implications. Energy Policy 37(11):4208–4219. https://doi.org/10.1016/j.enpol.2009.05.020

Das A, Parikh J (2004) Transport scenarios in two metropolitan cities in India: Delhi and Mumbai. Energy Convers Manage 45(15–16):2603–2625. https://doi.org/10.1016/j.enconman.2003.08.019

Femenías P, Fudge C (2010) Retrofitting the city: reuse of non-domestic buildings. Urban Des Plan 163(3):117–126. https://doi.org/10.1680/udap.2010.163.3.117

Gartland L (2008) Heat island: understanding and mitigating heat in urban areas. Earthscan, London. https://doi.org/10.4324/9781849771559

Golubchikov O (2004) Urban planning in Russia: towards the market. Eur Plan Stud 12(2):229–247. https://doi.org/10.1080/0965431042000183950

Golubchikov O (2011) Climate neutral cities: how to make cities less energy and carbon intensive and more resilient to climatic challenges. United Nations Economic Commission for Europe (UNECE), Geneva, p 85. https://www.unece.org/fileadmin/DAM/hlm/documents/Publications/climate.neutral.cities_e.pdf

Goodier C, Rydin Y (2010) Sustainable energy and sustainable cities. Urban Des Plan 163(4):147–148. https://doi.org/10.1680/udap.2010.163.4.147

Hart D (2003) Hydrogen—a truly sustainable transport fuel? Front Ecol Environ 1(3):138–145. https://doi.org/10.1890/1540-9295(2003)001%5b0138:HATSTF%5d2.0.CO;2

Hopkins R (2010) What can communities do? In: Heinberg R, Lerch D (eds) The post carbon reader: managing the 21st century's sustainability crises. Watershed Media, Healdsburg, CA, pp 442–454

HTA, Levitt Bernstein, PRP, PTEarchitects, Design for Homes (2007) Recommendations for living at superdensity. Design for Homes, London, p 32. http://www.designforhomes.org/wp-content/uploads/2012/03/Superdensity2.pdf

Hunt JCR, Maslin M, Killeen T, Backlund P, Schellnhuber HJ (2007) Climate change and urban areas: research dialogue in a policy framework. Philos Trans R Soc 365(1860):2615–2629. https://doi.org/10.1098/rsta.2007.2089

IEA (2018) CO_2 emissions from fuel combustion, 2018 edn. International Energy Agency (IEA), Paris, p 515. https://doi.org/10.1787/co2_fuel-2018-en

James T (2009) Built to be green. Eng Technol 4(20):52–53. https://doi.org/10.1049/et.2009.2011

Jelle BP, Gustavsen A Baetens R (2010) The path to the high performance thermal building insulation materials and solutions of tomorrow. J Build Phys 34(2):99–123. https://doi.org/10.1177/1744259110372782

Jiang P, Tovey K (2010) Overcoming barriers to implementation of carbon reduction strategies in large commercial buildings in China. Build Environ 45(4):856–864. https://doi.org/10.1016/j.buildenv.2009.09.004

Jiang H, Chen FJ, Wang QM (2010) Forest carbon sinks information acquisition and regional low carbon development analysis. In: IEEE Xplore (20 Jan 2011), 3rd international conference on information management, innovation management and industrial engineering, Kunming, China, 26–28 Nov 2010, pp 409–412. https://doi.org/10.1109/ICIII.2010.418

Jing B, Sun Z, Liu M (2010) China's energy development strategy under the low carbon economy. Energy 35(11):4257–4264. https://doi.org/10.1016/j.energy.2009.12.040

Jones A (2009) Moving cities towards a sustainable low carbon energy future: learning from working and London. In: Davoudi S, Crawford J, Mehmood A (eds) Planning for climate change: strategies for mitigation and adaptation for spatial planners. Earthscan, London, pp 273–283. https://doi.org/10.4324/9781849770156

Kårstein A (2010) Stumbling blocks on the Hydrogen road in Norway. Urban Des Plan 163(4):177–183. https://doi.org/10.1680/udap.2010.163.4.177

Kelly MJ (2009) Forum: retrofitting the existing UK building stock. Build Res Inf 37(2):196–200. https://doi.org/10.1080/09613210802645924

Kemp M, Wexler J (eds) (2010) Zero-carbon Britain 2030: a new energy strategy. Centre for Alternative Technology, Llwyngwern, p 368. https://b.3cdn.net/nefoundation/7e8212826db20c2c3f_qcqm67257.pdf

Li J, Colombier M, Giraud PN (2009) Decision on optimal building energy efficiency standard in China—the case for Tianjin. Energy Policy 37(7):2546–2559. https://doi.org/10.1016/j.enpol.2009.01.014

Li L, Zhou H, Cao D (2010) A new model of china's urban construction—exploration of a low carbon city. In: IEEE Xplore (16 Sept 2010), international conference on management and service science, Wuhan, China, 24–26 Aug 2010, pp 1–4. https://doi.org/10.1109/ICMSS.2010.5577909

Lindsey M, Schofer JL, Durango-Cohen P, Gray KA (2011) The effect of residential location on vehicle miles to travel, energy consumption and greenhouse gas emissions: Chicago case study. Transp Res 16(1):1–9. https://doi.org/10.1016/j.trd.2010.08.004

Maassen A (2010) Planning urban energy trajectories: London and Barcelona. Urban Des Plan 163(4):185–192. https://doi.org/10.1680/udap.2010.163.4.185

McIntyre JH, Lubitz WD, Stiver WH (2011) Local wind-energy potential for the city of Guelph, Ontario (Canada). Renew Energy 36(5):1437–1446. https://doi.org/10.1016/j.renene.2010.10.020

Moloney S, Horne RE, Fien J (2010) Transitioning to low carbon communities—from behaviour change to systematic change: lessons from Australia. Energy Policy 38(12):7614–7623. https://doi.org/10.1016/j.enpol.2009.06.058

Nadin V (2006) Balanced urban growth, sustainability and the role of spatial planning. In: Aspects of equilibrium, Warsaw, Poland, 23–25 June 2005. Oficyna Wydawnicza Politechniki Wroclawskiej, Wroclaw, Poland, pp 220–221

Nilsson SF, Reidhav C, Lygnerud K, Werner S (2008) Sparse district-heating in Sweden. Appl Energy 85(7):564. https://doi.org/10.1016/j.apenergy.2007.07.011

Norman J, MacLean HL, Kennedy CA (2006) Comparing high and low residential density: life-cycle analysis of energy use and greenhouse gas emissions. J Urban Plan Dev 132(1). https://doi.org/10.1061/(ASCE)0733-9488(2006)132:1(10)

OECD (2010) Green cities programme. Organisation for Economic Co-operation and Development (OECD), Paris, p 4. https://www.oecd.org/regional/greening-cities-regions/46811501.pdf

Ou X, Zhang X, Chang S (2010) Alternative fuel buses currently in use in China: life-cycle fossil energy use, GHG emissions and policy recommendations. Energy Policy 38(1):406–418. https://doi.org/10.1016/j.enpol.2009.09.031

Owens SE (1986) Energy, planning, and urban form. Pion, London

Peltier R (2009) The Hague repowering project upgrades CHP system, preserves historic building. Power 153(8):38–44

Perrels A (2008) Sustainable mobility and urbanity. In: Perrels A, Himanen V, Lee-Gosselin M (eds) Building blocks for sustainable transport: obstacles, trends, solutions. Emerald, Bingley, pp 133–156

Phdungsilp A (2010) Integrated energy and carbon modelling with a decision support system: policy scenarios for low-carbon city development in Bangkok. Energy Policy 38(9):4808–4817. https://doi.org/10.1016/j.enpol.2009.10.026

Power A (2010) Housing and sustainability: demolition or refurbishment? Urban Des Plan 163(4):205–216. https://doi.org/10.1680/udap.2010.163.4.205

PRP Architects Ltd., URBED, Design for Homes (2008) Beyond eco-towns: applying the lessons from Europe. PRP Architects Ltd., London, p 29. http://urbed.coop/sites/default/files/Beyond%20Ecotowns.pdf

Rawlings P (2010) Climate friendly buildings and offices: a practical guide. United Nations Environment Programme, Nairobi, p 135. https://www.greeningtheblue.org/sites/default/files/climate-friendly-buildings-final_0.pdf

Ren H, Zhou W, Nakagami K, Gao W, Wu Q (2010) Feasibility assessment of introducing distributed energy resources in urban areas of China. Appl Therm Eng 30:2584–2593. https://doi.org/10.1016/j.applthermaleng.2010.07.009

Rice L (2010) Retrofitting suburbia: is the compact city feasible? Urban Des Plan 163(4):193–204. https://doi.org/10.1680/udap.2010.163.4.193

Roaf S, Crichton D, Nicol F (2009) Adapting buildings and cities for climate change: a 21st century survival guide, 2nd edn. Architectural Press, Elsevier Ltd., Oxford, p 385. http://library.uniteddiversity.coop/Ecological_Building/Adapting_Buildings_and_Cities_for_Climate_Change.pdf

Rydin Y (2010) Governing for sustainable urban development. Earthscan, London. https://doi.org/10.4324/9781849775083

Rydin Y, Devine-Wright P, Goodier C, Guy S, Hunt L, Watons J (2010) Briefing: challenging lock-in through urban energy systems. Urban Des Plan 163(4):149–151. https://doi.org/10.1680/udap. 2010.163.4.149

Satterthwaite D (2011) How urban societies can adapt to resource shortage and climate change. Philos Trans R Soc 369(1942):1762–1783. https://doi.org/10.1098/rsta.2010.0350

Shinnar R, Citro F (2006) A road map to U.S. decarbonisation. Science 313(5791):1243–1244. https://doi.org/10.1126/science.1130338

Stanilov K (ed) (2007) The post-socialist city: urban form and space transformation in Central and Eastern Europe after socialism. Springer, London, p 490. https://doi.org/10.1007/978-1-4020-6053-3

Steemers K (2003) Energy and the city: density, buildings and transport. Energy Build 35(1):3–14. https://doi.org/10.1016/S0378-7788(02)00075-0

Stevens MR, Berke PR, Song Y (2010) Creating disaster-resilient communities: evaluating the promise and performance of new urbanism. Landscape Urban Plan 94(2):105–115. https://doi.org/10.1016/j.landurbplan.2009.08.004

Sveum P, Kizilel R, Khader M, Al-Hallaj S (2007) IIT plug-in conversion project with the City of Chicago. In: IEEE Xplore (17 June 2008), 2007 IEEE vehicle power and propulsion conference, Arlington, TX, USA, 9–12 Sept 2007, pp 493–497. https://doi.org/10.1109/VPPC.2007.4544174

Thornbush M (2015) AIMS Environ Sci 2(3):852–867. https://doi.org/10.3934/environsci.2015.3. 852

Thornbush MJ, Viles HA (2007) Simulation of the dissolution of weathered versus unweathered limestone in carbonic acid solutions of varying strength. Earth Surf Proc Land 32(6):841–852. https://doi.org/10.1002/esp.1441

UN-Habitat (2010) Bridging the urban divide: why cities much build equality. Urban World 1(5):8–26. http://mirror.unhabitat.org/pmss/getElectronicVersion.aspx?nr=2880&alt=1

Ürge-Vorsatz D, Koeppel S, Mirasgedis S (2007) Appraisal of policy instruments for reducing buildings' CO_2 emissions. Build Res Inf 35(4):458–477. https://doi.org/10.1080/09613210701327384

Vandevyvere H, Stremke S (2012) Urban planning for a renewable energy future: methodological challenges and opportunities from a design perspective. Sustainability 4(6):1309–1328. https://doi.org/10.3390/su4061309

Wang K, Cao D (2010) A new path explore of city's digit management. In: IEEE Xplore (14 Oct 2010), 6th international conference on wireless communications networking and mobile computing, Chengdu, China, 23–25 Sept 2010, pp 1–4. https://doi.org/10.1109/WICOM.2010. 5601214

Williams K, Joynt JLR, Hopkins D (2010) Adapting to climate change in the compact city: the suburban challenge. Built Environ 36(1):105–115. https://doi.org/10.2148/benv.36.1.105

World Bank (2008) Climate resilient cities: a primer on reducing vulnerabilities to climate change impacts and strengthening disaster risk management in East Asian cities. The International Bank for Reconstruction and Development/The World Bank, Washington DC, p 157. https://siteresources.worldbank.org/INTEAPREGTOPURBDEV/Resources/Primer_e_book.pdf

Zannis G, Santamouris M, Geros V, Karatasou S, Pavlou K, Assimakopoulos MN (2006) Energy efficiency in retrofitted and new museum buildings in Europe. Int J Sustain Energ 25(3–4):199–213. https://doi.org/10.1080/14786450600921645

Zhang X, Ruoshui W, Molin H, Martinot E (2010) A study of the role played by renewable energies in China's sustainable energy supply. Energy 35(11):4392–4399. https://doi.org/10.1016/j.energy. 2009.05.030

Zhao H, Chu Y (2009) Study on urban design strategy for low carbon trip. In: IEEE Xplore (28 Dec 2009), international conference on energy and environment technology, 16–18 Oct 2009, pp 373–377. https://doi.org/10.1109/ICEET.2009.96

Chapter 4
Becoming Smart

Abstract Based on the strategies of smart cities from around the world, an initial study and results are relayed here based on a sample of 30 cities. Subsequently, more case studies were added to the roster representing 50 strategies. A selection of 10 studies was then identified for a more in-depth focus on actually existing case studies. This chapter conveys a diversity of cases based on actually existing plans for smart development based on smart strategies. In this way, it is possible to pinpoint 'actually existing smart cities' from around the world. Inherent in the unique cases is a sense of the disparate priorities evident in smart strategies and cases contingent on their location.

Keywords Actually existing smart cities · Smart strategies · Case studies/cases · Roster

In the second portion of this brief, the focus will be on actually existing smart cities and how they convey a progression from low carbon transitions—advocating reduced resource consumption and waste—to become smart cities. The aim is to identify actually existing case studies from around the world (in both developed and developing countries) based on a foreseen total of 50 case studies, with 10 in-depth examples following from others, such as Anthopoulos (2017). A case study roster containing information of city projects, including year, rationale, objectives, and website address was developed to inform the analysis based on strategic reports and website content, plus publications. In particular, the authors were interested in self-defined smart city 'strategies', which varied based on divergent approaches evident in actually existing smart cities.

An initial analysis based on 30 cities already revealed some trends (denoted in square brackets under cities in Table 4.1). The trends convey a concentration of smart city strategies or plans in Europe and Asia, followed by North America, then the developing world that includes countries in South America and Africa. Australia (and Oceania) had the lowest count among all continents (similar to Lee & Hancock 2012, conveyed in square brackets for comparison under continents), with South America also having the lowest count among the 30 strategies identified in the initial study.

© The Author(s), under exclusive license to Springer Nature Switzerland AG 2020   35
M. J. Thornbush and O. Golubchikov, *Sustainable Urbanism in Digital Transitions*, SpringerBriefs in Geography,
https://doi.org/10.1007/978-3-030-25947-1_4

Table 4.1 Initial findings based on a sample of 30 cities compared with Lee and Hancock (2012; summarised in the continents column) based on 143 smart green city projects

Continents	Cities	Years
Asia [40]—Lee and Hancock (2012)	Dubai (UAE), Hong Kong (China), Pune (India), **Seoul (South Korea)**, Singapore, Taipei (Taiwan), Tel Aviv (Israel), Yinchuan (China) [8]	2011–2014
North America [35]	Boston (USA), Chicago (USA), Edmonton Canada), New York (USA), Washington, DC (USA) [5]	2010–2015
Europe [47]	Barcelona (Spain), Bologna (Italy), Eindhoven (Netherlands), Gothenburg (Sweden), Lisbon (Portugal), Manchester (UK), Milan (Italy), Milton Keynes (UK), Moscow (Russia), Rijeka (Croatia), Stockholm (Sweden), **Tampere (Finland)**, Vienna (Austria) [13]	2008–2020
South America [11]	Rio de Janeiro (Brazil) [1]	2010–2016
Middle East and Africa [10]	South Africa (e.g., Cape Town) [2]	2000–2014
Australia (Oceania) [7]	Melbourne (Australia) [1]	2001–2004

In addition to the spatiality of the strategies for smart development, there is a temporal framework apparent that follows the results by Lee and Hancock (2012). The earliest strategies appeared from the 2000s, as for example 2008 for Europe (Tampere, Finland) and 2011 for Asia (Seoul, South Korea), which are emboldened in Table 4.1. It is, therefore, not surprising that most smart cities are located on the European continent, followed by Asia, given that these have had the longest investment in smart development. Consequently, there are currently more smart cities located in the Global North than in the Global South. Among developed countries, clusters appear in North America (east coast) and Europe.

These findings from the initial analysis support the results by Lee and Hancock (2012), based on 143 cities, noted in Table 4.1. Their research notes the following aims of smart cities: '… to implement smart technologies to address and resolve such urban problems such as energy shortages, traffic congestion, inadequate urban infrastructure, and some issues in health and education. In particular, the European Union (EU) is investing in smart city strategies for metropolitan city regions such as Barcelona, Amsterdam, Berlin, Manchester, Edinburgh and Bath'. City regions add to observations made earlier (in Chap. 1) regarding scale-wise growth, spreading from the city-scale to embrace entire regions and beyond.

Frost and Sullivan (2013; also Glasmeier & Christopherson 2015 for the propagation of the concept since the 1980s) relay research on smart city projects and initiatives. Based on key parallels across these schemes, eight key aspects of the smart city were identified and used to define it, which included: smart governance and education, healthcare, building, mobility, infrastructure, technology, energy, and

Table 4.2 Global smart cities (26) by 2025 according to Frost and Sullivan (2013)

Country/continent	City
Canada (3)	Vancouver[+], Calgary, Toronto[+]
United States (5 + 2[*])	Seattle[+], San Francisco, Los Angeles, San Diego[*], Chicago, Boston[*], New York[+]
Europe (10 + 2[*])	Glasgow[*], London (UK)[+]; Barcelona (Spain)[+]; Amsterdam (the Netherlands)[+]; Luxembourg[*]; Vienna (Austria)[+]; Berlin (Germany)[+]; Copenhagen (Denmark); Oslo (Norway)[+]; Stockholm (Sweden)[+]; Helsinki (Finland)[+]
Africa (0 + 1[*])	Johannesburg[*] (South Africa)
Asia (7 + 3[*])	Delhi[*] (India); Chengdu[*], Wuhan, Shenzhen, Beijing, Tianjin (China); Seoul (South Korea)[+]; Tokyo (Japan); Singapore (Singapore)[+]; Jakarta[*] (Indonesia)
Australia (1)	Sydney[+]

[*]Represents (8) small-scale projects and/or specific initiatives not included in their projected 26 global smart cities
[+]Indicate overlap with cities included in the roster of the current study

citizen. The firm narrowed its definition to encompass five out of the eight smart parameters, as none existed (at the time) that had all eight attributes. Those that did not have at least five of the parameters were considered to be 'eco-friendly' cities, as for example, Nice, France. Furthermore, projects that were too small because they were single developments and did not entail entire cities (e.g. Masdar City, also Song Do in South Korean, PlanIT in Portugal; Glasmeier & Christopherson 2015; also relayed as two pilots, relevant for technological learning and societal embedding from a sociotechnical perspective; Carvalho 2015) were also excluded from their classification. It was estimated that by 2025 there would be in existence at least 26 global smart cities—that had at least five of his eight parameters—actually in existence. Of these, half were expected to be located in Europe and North America.

All of these cities would be different, emphasising some parameters over others—for example, Amsterdam was expected to execute projects in governance, mobility, energy, and so on, with funding coming from the Amsterdam City Project, including city government and private participants as well as the EU. It was also projected that cities such as in Spain would have sufficiently large GDPs to contribute to a smart city market of c. $1.5 trillion (globally) to affect various sectors, including governance, healthcare, building, transportation, infrastructure, and energy (Frost & Sullivan 2013). Smart cities projected to be included for 2025 are listed in Table 4.2. Finally, the firm recognised four main key players in the smart city, including integrators (platforms), network service providers (collaborative networks, data analytics, enterprise), pure-play product vendors (hard assets, e.g. smart metres, distribution devices, etc.), and managed service providers (monitoring, management, consulting).

Although most of such surveys for smart cities rely on authors' preconceptions on where to search for—and consequently where to find—the evidence of emerging smart cities and, while they also tend to be biased to larger cities and those regions and cities that have articulated their ambitions in the English language, we believe that these results, with a degree of approximation, are still indicative of the trend of greater representation of smart cities in developed countries (in the Global North).

Taking this proviso above seriously, it is still interesting to consider Juniper Research (2018; on behalf of Intel—a technology company-based in Silicon Valley, for research performed in 2015) that identified the 20 smartest or 'cleverest' cities. This ranking was established based on various variables, including cost of living, liveability, career opportunities, pollution, crime, and more, facilitated by digital technology (connected technology, including sensor, meters, lights, and so on to collect and analyse data to improve public infrastructure and services), shared knowledge, and social cohesivity—measured through mobility, public safety, health, and productivity. According to this market research organisation, the following rank was derived for its Smart City Index 2017 of 20 cities—some (*) overlapping with the research roster in the initial study:

1. Singapore, Singapore*
2. London, UK*
3. New York, USA*
4. San Francisco, USA
5. Chicago, USA
6. Seoul, South Korea*
7. Berlin, Germany*
8. Tokyo, Japan
9. Barcelona, Spain*
10. Melbourne, Australia*
11. Dubai, UAE*
12. Portland, USA
13. Nice, France
14. San Diego, USA
15. Rio de Janeiro, Brazil*
16. Mexico City, Mexico
17. Wuxi, China
18. Yinchuan, China
19. Bhubaneswar, India
20. Hangzhou, China

These cities are argued in the report to be able to 'give back' 125 hours per year to every resident, saving them time. Across four key areas (mobility, healthcare, public safety, productivity), these smart cities increase time savings as well as productivity and overall quality of life, increased health, and a safer environment in which to live (Juniper Research 2018). The research has revealed that an Internet of Things/IoT-integrated infrastructure based on intelligent traffic systems, for example, with directed parking, frictionless toll and parking payments, and safer roads, can itself 'give back' up to 60 hours per year because road congestion and gridlock create inefficiency in mobility.

Public transportation is indeed a key service in cities, with the capability of connecting citizens to an intelligent management strategy (e.g. Ramaswami et al. 2016). According to them, infrastructure capable of extending beyond the local to embrace transboundary scales is essential for the reach of smart cities. New York City, for example, is well-positioned because of its extensive public transportation system that has a high ridership, with 54% of commuters taking public transit (including bus/trolley bus, streetcar/trolley car, subway, railroad, ferryboat; Loo et al. 2010, based on The American Community Survey 2005 retrieved on 10 November 2008 from http://www.census.gov/acs/www/). Bike-sharing systems have been deployed in New York City (Noland et al. 2016), with findings suggesting that the placement of bikeshare stations is crucial—so that, when placed near busy locations (e.g. subway stations) and where there is bicycle infrastructure, there is greater use of such schemes. Furthermore, population size and employment also affect usage, with residential population representing more trips throughout the week, especially on non-working days.

New technologies are deployed, including Apps, that enable connectivity—as through IoT that connects devices over the Internet—and ease of usage/access, as through automated fare collection and vehicle location systems to plan journeys (Liu et al. 2017). According to these authors, the three pillars on which smart cities rest are data mining technology, IoT, and mobile wireless networks.

It is interesting to observe that many—a quarter—of the listed cities are based in the USA (New York, San Francisco, Chicago, Portland, San Diego—comprise five of the 20 listed cities). Following this, European countries (the UK, Germany, Spain, France) and China (Wuxi, Yinchuan, Hangzhou) have the greatest number of cities represented in the ranking. Singly-represented cities are from the remaining countries, including Singapore (ranked first in the world), South Korea for Seoul, Japan for Tokyo, UAE for Dubai, Brazil for Rio de Janeiro, Mexico for Mexico City, and India for Bhubaneswar. Only three countries (Singapore, Australia, Brazil) located in the Global South are included in this ranking, with the remaining majority of smart cities coming from developed countries located in the Global North. African countries, for instance, are not represented in such research. Again, this probably also indicates a problem with methodology, where only available and self-promoted resources from cities become part of such rankings' consideration.

In our search, we wanted to be more inclusive and yet focused. However, we were also linguistically-constrained and relied on the Internet as a source for information. In what follows, we present a detailed overview of 10 cities selected based on their smart strategy. There is a significant overlap with these ranked cities and those chosen as detailed cases in this brief.

4.1   Strategies Roster

The roster of strategies compiled for this research identifies a 'strategy' as a city plan for smart development. This is broadly defined and includes projects and initiatives across sectors as well as (mayoral) strategies available online in English in PDF format from various websites, some pertaining to specific smart city strategies. An initial dataset of case studies was compiled that considered a variety of projects, including year(s) of operation, purpose or rationale, and website for reference.

A summary of roster themes/attributes—based on a comparison of priorities and initiatives—appears in Table 4.3. It considers fewer than 100 cities, as deployed by others like Broto and Bulkeley (2013) for a database encompassing five key sectors (urban infrastructure, built environment, transport, carbon sequestration, urban form) of climate change mitigation. However, the current selection of 50 represents actually existing smart cities—based on English online PDFs of smart strategies available for each city. Shelton et al. (2015) similarly advocate the 'actually existing smart city' as opposed to paradigmatic smart cities (e.g. Songdo, Masdar, Living PlanIT Valley). They opt to focus on cases in Louisville and Philadelphia in the USA from where to consider the impacts of policies on actual cities around the world.

Table 4.3 Summary of smart strategy themes/attributes based on roster information

Common themes	Unique attributes
Accessibility, e.g. elderly/computer literacy; Open Access	Commercialisation—Edmonton
Autonomous vehicles	Commodification—Toronto
Cost reduction	Democratic right/data protection—Berlin
Digital government and services	Dual use, e.g. lamp posts and electric vehicle (EV) charging—Leipzig
Digital inclusion/bridging digital divide	eHealth services—London (NHS)
Ecology/environment; reduce carbon dioxide (CO_2) emissions	Footprint—Stockholm
Efficiency	Hackathons—Melbourne
Energy/energy grids/electric vehicles (EVs)/green energy/renewable energy use	Happy citizens and well-being—Dubai, Hong Kong
Interconnectedness, e.g. free Wi-Fi, platforms	Hubs—Leipzig, Milton Keynes, Taiwan
Mayor-led strategies—e.g. London	Low-cost—Pune, Stockholm, Tel Aviv
Partnerships/collaborations, e.g. academics/universities; citizen-centred partnerships	Retrofitting—Oslo
Shared services and information, e.g. bikeshares, Big Data	Structures, e.g. shelters, meltways, etc.—Toronto
Spur economy	Tech-first approach—Chicago
Start-ups and entrepreneurship	Tube system of waste disposal—Songdo
Sustainability	Unified/interdisciplinary approach/holistic perspective—Dallas, Heraklion, Lisbon
–	Urban greening and farming—South Korea
–	Vulnerable, e.g. elderly—Tshwane

Based on Table 4.3, it is evident that many actually existing smart cities denote unique attributes in their strategies, with some examples provided in the table for illustration (they are, of course, by far not all-encompassing). On the other hand, common themes also emerged from the roster of 50 cities from around the world in this study, conveying priorities and initiatives concerning accessibility, reduced cost, improved services, inclusion, improved environment, energy efficiency, renewable energy, shared services and information, aims to kickstart or spur the economy—as through innovation and business start-ups/entrepreneurship, and an emphasis on (up to nine, but commonly five or six) components of sustainability (social, economic, environmental, etc. as pillars).

4.2 Detailed Cases

Detailed contemporary and known case studies outline smart strategies for a selection of 10 'actually existing' cases of smart cities and initiatives. This information is contained in Table 4.4. Authors, such as Carvalho (2015), have already adopted a case study approach to compare advancements in smart cities—as for instance his comparison of Songdo (Incheon, South Korea) as the ubiquitous-city or 'u-city' to test new technologies and PlanIT Valley in Portugal as the 'city with a brain' and with a research and development/R&D focus (Paredes, 30 km from Porto) from a sociotechnical perspective. As outlined by the author, the former was built up from land that was reclaimed from the sea and in the latter case on a greenfield site (see his Table 1, p 49). Both smart cities have operating systems, but with different providers (e.g. Cisco for Songdo), and a central operation centre for Songdo versus an Urban Operating System deployed in PlanIT Valley (along with sensors and urban Apps). Whereas both local and national government drove Songdo, PlanIT Valley was constructed by an IT company (Living PlanIT), with local and national governments supporting the initiative. Evidently, smart cities vary in their drivers, and it is important to examine the rationale of the strategies in place to develop them. The 10 smart cities with coherent strategies are detailed in Table 4.4.

As noted by Odendaal (2006), one of the main challenges limiting smart city development is overcoming the digital divide. In Cape Town (South Africa), for instance, a digital divide assessment identified technical access, education and training, affordability, and sociocultural factors as constraining smart cities. These factors are surely issues affecting other, particularly low-income, countries. In the case studies presented in Table 4.4, it is evident that developing countries are severely under-represented, indicating that they do not have developed strategies in place for smart city development (or at least that these strategies are not widely available to the public). Lee and Hancock (2012), for instance, identify 143 smart green city projects already in existence before 2013, among them 35 in North America, 11 in South America, 47 in Europe, 40 in Asia, and 10 in the Middle East and Africa (plus seven in Australia and Oceania). So, there is an abundance of smart city development among European cities, and this is reflected in the literature as well as the roster presented here.

Table 4.4 Detailed case studies of a selection of 10 cities with recent or known smart strategies (listed alphabetically by city)

City (Country), year(s)—strategy details	Case highlights
Berlin (Germany), 2015 to 2030—Smart City Strategy Berlin. Available from: https://www.berlin-partner.de/fileadmin/user_upload/01_chefredaktion/02_pdf/02_navi/21/Strategie_Smart_City_Berlin_en.pdf	– Six areas of action: smart administration and urban society, housing, economy, mobility, infrastructure, and public safety—latter is the most relevant to citizenship; also recognises public democratic right to data protection – Adopts an integrated (ecological, social, economic, cultural) approach to finding solutions to challenges, involving actors from politics, business, science, administration, and urban society
Brussels (Belgium), 2014 to 2019—The Brussels Smart City Strategy. Available from: https://s3.eu-west-1.amazonaws.com/expopolis-scve/Fair1/TheBrusselsSmartCitystrategy.pdf	– Five dimensions to digital society and economy: connectivity, human capital, Internet use and the digital divide, more effective public services—opening up data (Open Access) and data analysis (Big Data), and digital public services – Sustainable development responding to ecological issues, including a governance model advocating participation and collaboration; digital inclusion
Cape Town (South Africa), 2001 to 2005—Cape Town's "Smart City" Strategy in South Africa. Available from: http://unpan1.un.org/intradoc/groups/public/documents/cpsi/unpan033820.pdf	– Five pillars: leadership in technology policy and strategy, economic and social development, digital democracy, more efficient and effective local government—with reduced transaction costs, and anytime-anywhere citizen services – Vision where citizens are connected to each other and the world; residents have access to ICT, having the skills to use it, bridging the digital divide
Columbus (Ohio, USA), 2012 to 2050—Columbus Smart City Application. Available from: https://www.eenews.net/assets/2016/03/31/document_pm_02.pdf	– Approaches' challenges by embracing existing infrastructure, networks, and data. Facing issues of an ageing population, youthful urban areas, mobility challenges, and a growing economy and population—that has housing, commercial, and passenger/freight, and environmental issues – Five strategies: access to jobs, smart logistics, connected visitors, connected citizens, and sustainable transportation – Vision for beautiful, healthy, and prosperous city. Major and public/private cooperation through Columbus Partnership
Edmonton (Alberta, Canada), 2009 to 2040— Smart City Strategy 2017. Available from: https://www.edmonton.ca/city_government/documents/PDF/Smart_City_Strategy.pdf	– Six 10-year strategic goals based on the main streams of resiliency, liveability, and workability – Aims to increase funding to local health and accelerate commercialisation of new technologies and products; also smart transportation – The Way Ahead—based on citizen-built vision, including efforts to deliver the greatest value of services and infrastructure

(continued)

Table 4.4 (continued)

City (Country), year(s)—strategy details	Case highlights
Milano (Italy), 2015 to 2020—Guidelines—Milano Smart City. Available from: http://www.milanosmartcity.org/joomla/images/sampledata/programma/SmartCity/milano%20smart%20city%20-%20guidelines.pdf	– Aims to achieve energy smartness at the city level to reach CO_2 reduction target by 2020 – Ten targets: deploy smart city solutions and accelerate their uptake, innovative models, external investment, energy-efficient districts, local renewable energy sources, new models of e-mobility, engage with citizens, exploit city data, and foster innovation locally for the creation of new businesses and jobs
Milton Keynes (England, UK), 2014 to 2017—http://www.mksmart.org/: https://www.milton-keynes.gov.uk/assets/attach/51579/Milton%20Keynes%20Digital%20Strategy%202018-2025.pdf MK Digital Strategy 2018–2025 by Milton Keynes Council. Available from: https://www.milton-keynes.gov.uk/assets/attach/51579/Milton%20Keynes%20Digital%20Strategy%202018-2025.pdf	– Ambitions to becoming an energy-efficient city, with reduced carbon emissions. Has installations, e.g. Falcon smart grid, EV charging infrastructure, and district heating system – MK: Smart has two main goals: innovative energy services based on the capabilities of MK Data Hub and demonstrate its business value to the energy sector – Exploring ways to manage water, energy, and transport—in this fast-growing city—using Big Data – MK Digital Strategy prioritises digital connectivity, services, and economy; aspires to be collaborative, innovative, and inclusive—e.g. improving access
Moscow (Russia), 2018 to 2030—'Moscow Smart City—2030': A Brief Version. Available from: https://2030.mos.ru/netcat_files/userfiles/documents_2030/strategy_tezis_en.pdf	– Mission seeks to improve city environment, create a favourable business environment, and improve living standards and city governance efficiency and transparency—based on Big Data and artificial intelligence (AI), consolidated society, enhanced active life for the elderly, and improved citizen well-being – Focuses on: human and social resources, urban environment, digital mobility, city economy, safety and ecology, and digital government
Taipei (Taiwan), 2017—The Implementation of Taipei Smart City Project by W.-B. Lee. Available from: https://www.lct.tp.edu.tw/ezfiles/1/1001/img/207/110629308.pdf	– Promotes collaboration between public and private sectors. Aims to become a hub for smart city industry – Seeks to strengthen national broadband network, e.g. by expanding Wi-Fi access on Taiwan High Speed Rail network—passengers able to access the Internet on all trains using an iTaiwan account, and cultivate innovative services and talent
Tel Aviv (Israel), 2016—Tel Aviv Smart City. Available from: https://www.tel-aviv.gov.il/en/WorkAndStudy/Documents/SMART%20CITY%20TEL%20AVIV.pdf	– Focuses on direct resident-oriented (lightweight) services; decentralised, low-cost methods—for a cost-effective smart city initiative – Reliance on local start-up ecosystem, creating services, using open municipal databanks, and public–private partnerships

4.3 Evaluation

Smart cities are important considerations in the years to come from a multitude of approaches and disciplinary frameworks, including multidisciplinary approaches that address different components of this facet of the built environment. According to Causone et al. (2017), smart cities represent a hub in global energy-flow networks and, as such, are an essential element of the environment affected by urbanisation. Causone et al. (2017, p. 868) provide figures for contemporary urban growth: 'Currently, 75% of EU and 81% of the US population already lives in urban areas, whereas, according to a UN report, the largest urban growth by 2050 will take place in Asia and Africa'—see, for instance, the World City Populations Interactive Map 1950–2030 (Luminocity3d.org. 2016), designed by DA Smith at CASA UCL with UN data from 2014. However, much of the smart city development identified in the contemporary literature and strategies considered in this study have been for the Global North, with Asia experiencing some growth but still lagging behind North America and Europe. Africa has had very little smart city growth until very recently, and will lag behind the rest of the world—even though it is a continent that is expected to experience continued urbanisation in the decades to come (cf. Odendaal 2006). However, it is expected that African countries will follow suit and their smart development shall intensify in the next decades.

The current study has focused on self-defined smart cities based on strategies that involve projects and initiatives making up smart development. The emphasis has been on the actually existing smart city (after e.g. Shelton et al. 2015), and a neoliberal approach has been supported in terms of piecemeal development based on digital technologies. Authors such as Cho (2017) have captured how surveillance and data mass capture indicate the importance of being tracked by digital technology that can convey spatial data. Digital data can also relay anytime-anywhere information, providing access and control over people's movements (Cosgrave et al. 2013). According to the authors, this has created 'information marketplaces' that point to the potential for commodified information, with implications for national private security.

It is integral that the implications of smart cities continue to be considered in the literature by academics as well as decision-makers and policymakers. It is not enough to evaluate the smart city in isolation for specific criteria; instead, holistic approaches are preferable, as they may capture a broader perspective than current piecemeal development. What are the implications of smart technology, including that based on AI, for the present but also the future; what could be the potential outcomes? It is easy to get carried away with technological advancement and, in the delivery, forget its ultimate (perhaps the initial) purpose. Robotics, for example, will be considered later in the final chapter of this brief. Here, however, it is necessary to evaluate the future of smart development. Because of its piecemeal evolution from an entrepreneurial approach geared towards making cities more competitive, the overall direction and implications are necessary to consider as part of the challenge.

It has been stated, as for instance by Caird (2018)—who conducted case study-based research in the English cities of Birmingham, Bristol, Manchester, Milton Keynes, and Peterborough—that more work is needed to evaluate smart interventions for both cities and citizens. A sociotechnical approach enables for such a holistic or integrated assessment that considers technological issues as well as social issues and is, therefore, well-positioned to deal with 'wicked' urban problems (cf. Vasseur et al. 2017; see Chap. 6), as those stemming from socioeconomics. Similarly, addressing the British cities of Bristol, Manchester, Milton Keynes, and Peterborough—with Glasgow and London added over Caird's (2018) work—Cowley et al. (2018) consider smart city programmes from a techno-public approach. Using this perspective, they also discovered a dominating entrepreneurial (also service-user) public mode that implicated both civic and political roles in their assemblage of (six) smart cities. Joss (2018) specifically—for London, UK—addressed the technocentric view that has been increasingly challenged through a focus on the urban citizen in smart development. However, according to Joss, UK governance can circumscribe the planning and decision-making processes involved to limit public accountability. He has espoused that citizens adopt an entrepreneurial role as producers of information, thereby calling for the greater involvement (or placement) of people in the making of the cities of tomorrow. By so doing, he effectively contributes to a public governance focus as a common (reoccurring) trend in the contemporary research British agenda concerning smart city development.

Perhaps the greatest challenge to citizen involvement in smart cities is that alluded to by Glasmeier and Christopherson (2015) as that of inclusiveness when accessing digital technologies. This continues to be a challenge today because of the increasing wage divide and poverty, even in developed nations, restricting equal opportunities of access to technology. There is also the problem of ageing populations in developed countries around the world and skills challenges—connected with age and training/education—that can limit access to some demographics. Batty et al. (2012), for instance, present (six) research challenges that include the development of equitable technologies to improve the quality of life in cities and that enable informed (including online and mobile forms of) participation and knowledge-sharing as part of democratic governance. In addition, Kitchin (2015) refers to other (research-related) shortcomings concerning the smart city agenda, raising four problematic areas, including: (1) a lack of detailed genealogies and comparative research; (2) the use of canonical examples; (3) with an absence of empirical case studies of initiatives, plus 'one-size fits all narratives'; and (4) undeveloped partnerships with different stakeholders. The current research contributes to rectifying (2) and (3), in particular, by compiling a roster of actually existing case studies for an 'empirical' perspective as well as brining in a selection of detailed case studies—that can be employed for comparisons, so contributing to (1); however, (4) remains amiss here, except for advocating for interdisciplinary partnerships that also include academics.

References

Anthopoulos L (2017) Smart utopia VS smart reality: learning by experience from 10 smart city cases. Cities 63:128–148. https://dx.doi.org/10.1016/j.cities.2016.10.005

Batty M, Axhausen KW, Giannotti F, Pozdnoukhov A, Bazzani A, Wachowicz M, Ouzounis G, Portugali Y (2012) Smart cities of the future. Eur Phys J Spec Top 214:481–518. https://doi.org/10.1140/epjst/e2012-01703-3

Broto VC, Bulkeley H (2013) A survey of urban climate change experiments in 100 cities. Global Environ Change 23:92–102. https://doi.org/10.1016/j.gloenvcha.2012.07.005

Caird S (2018) City approaches to smart city evaluation and reporting: case studies in the United Kingdom. Urban Res Pract 11(2):159–179. https://doi.org/10.1080/17535069.2017.1317828

Carvalho L (2015) Smart cities from scratch? A socio-technical perspective. Camb J Reg Econ Soc 8:43–60. https://doi.org/10.1093/cjres/rsu010

Causone F, Sangelli A, Pagliano L, Carlucci S (2017) An exergy analysis for Milano smart city. Energy Procedia 111:867–876. https://doi.org/10.1016/j.egypro.2017.03.249

Cho L (2017) Mass capture: the making of non-citizens and the Mainland Travel Permit for Hong Kong and Macau residents. Mobilities 12(2):188–198. https://doi.org/10.1080/17450101.2017.1292776

Cosgrave E, Arbuthnot K, Tryfonas T (2013) Living labs, innovation districts and information marketplaces: a systems approach for Smart Cities. Procedia Comput Sci 16:668–677. https://doi.org/10.1016/j.procs.2013.01.070

Cowley R, Joss S, Dayot Y (2018) The smart city and its publics: insights from across six UK cities. Urban Res Pract 11(1):53–77. https://doi.org/10.1080/17535069.2017.1293150

Frost & Sullivan (2013) Strategic opportunity analysis of the global Smart City market, p 18. https://dsimg.ubm-us.net/envelope/153353/295862/1391029790_strategic_opportunity.pdf

Glasmeier A, Christopherson S (2015) Thinking about smart cities. Camb J Reg Econ Soc 8(1):3–12. https://doi.org/10.1093/cjres/rsu034

Joss S (2018) Future cities: asserting public governance. Palgrave Commun 4(1):36 (p 4). https://dx.doi.org/10.1057/s41599-018-0087-7

Juniper Research (2018) Smart cities—what's in it for citizens?, p 24. https://newsroom.intel.com/wp-content/uploads/sites/11/2018/03/smart-cities-whats-in-it-for-citizens.pdf

Kitchin R (2015) Making sense of smart cities: addressing present shortcomings. Camb J Reg Econ Soc 8:131–136. https://doi.org/10.1093/cjres/rsu027

Lee JH, Hancock M (2012) Toward a framework for smart cities: a comparison of Seoul, San Francisco and Amsterdam. In: Smart green cities conference on innovations for smart green cities: what's working, what's not, what's next, Stanford Business School, Palo Alto, USA, 26–27 June 2012

Liu Y, Weng X, Wan J, Yue X, Song H, Vasilakos AV (2017) Exploring data validity in transportation systems for smart cities. IEEE Commun Mag 55(5):26–33. https://doi.org/10.1109/MCOM.2017.1600240

Loo BPY, Chen C, Chan ETH (2010) Rail-based transit-oriented development: lessons from New York City and Hong Kong. Landscape Urban Plan 97:202–212. https://doi.org/10.1016/j.landurbplan.2010.06.002

Luminocity3d.org (2016) World city populations interactive map 1950–2030. Designed by DA Smith at CASA UCL with UN data from 2014. Available from http://luminocity3d.org/WorldCity/#3/23.40/10.90

Noland RB, Smart MJ, Guo Z (2016) Bikeshare trip generation in New York City. Transp Res A Policy 94:164–181. https://doi.org/10.1016/j.tra.2016.08.030

Odendaal N (2006) Towards the digital city in South Africa: issues and constraints. J Urban Technol 13(3):29–48. https://doi.org/10.1080/10630730601145997

Ramaswami A, Russell AG, Culligan PJ, Sharma KR, Kumar E (2016) Meta-principles for developing smart, sustainable, and healthy cities. Science 352(6288):940–943. https://doi.org/10.1126/science.aaf7160

Shelton T, Zook M, Wiig A (2015) The 'actually existing smart city'. Camb J Reg Econ Soc 8:13–25. https://doi.org/10.1093/cjres/rsu026

Vasseur L, Horning D, Thornbush M, Cohen-Shacham E, Andrade A, Barrow E, Edwards SR, Wit P, Jones M (2017) Complex problems and unchallenged solutions: bringing ecosystem governance to the forefront of the UN sustainable development goals. Ambio 46(7):731–742. https://doi.org/10.1007/s13280-017-0918-6

Chapter 5
Sociotechnical Issues

Abstract The influence of technology in smart cities is inevitable and continues to emerge from an entrepreneurial approach stemming from the business model. Both hardware and software components of technology are part of technical advancements in technologically advanced cities. Although more work has been published on smart cities, especially since the early 2010s, there remain uncertainties and challenges posed by smart cities that, in practice, could pose problems for society. This chapter addresses some of these social issues, including democratic governance opposed by monitoring and control in these technocratic (rather than democratic) cities. Security is addressed as Big Data and Open Access information amasses on the Internet and can be accessed worldwide. This represents one of the key areas, especially with the diffusion of the public-private boundary caused by continued monitoring and the accumulation of information on people, their movements and behaviours.

Keywords Technology · Economic/entrepreneurial approach · Governance · Democracy · Surveillance · Cybersecurity · Big Data · Internet of things/IoT · Networks · Social engagement

The emphasis on technology as providing solutions to growing cities is a double-edged sword. On the one hand, technology is capable of organising cities and ensuring that energy expenditure is controlled, so that efficient cities are preferred due to savings on resources—from an environmental business model. On the other hand, technology costs and not all cities around the world can afford to invest in advanced technology to solve their problems of rapid urban growth, environmental degradation, and socioeconomic issues. A technological fix cannot remedy all issues facing cities, although they can provide a means of enhanced efficiency, monitoring, and control.

In a previous publication, Thornbush et al. (2013) examine the sociotechnical dimension to urbanism, including the potential of cities to their reduce energy demand either through a technological approach, as by improving building energy performance or alternatively through behaviour change, as by reducing the need for motor vehicle use (see their Table 1, p 4). The authors also espouse (in Table 2, p 6) potential ways towards achieving low carbon urbanism, conveying social-technical dimensions that include the technical dimension, which incorporates urban energy infrastructure; building, urban design, and planning; and urban transport. Building on this study, it

© The Author(s), under exclusive license to Springer Nature Switzerland AG 2020   49
M. J. Thornbush and O. Golubchikov, *Sustainable Urbanism in Digital Transitions*, SpringerBriefs in Geography, https://doi.org/10.1007/978-3-030-25947-1_5

is possible to discern a common trend towards an integrated approach, so that smart cities actually encompass more than just components and rather aim to be integrative, targeting all aspects of these dimensions—including the human (social) dimension.

As addressed by other authors, such as Rose (2017), digitally-mediated urban spaces rely on software and digital hardware that operate as a technological non-human entity at the cost of human agency in what she terms 'post-human agency'. Sociotechnical agency can be spatial-temporally differentiated, according to her, in the way that it is organised as both diverse and innovative. As such, people can connect with such post-human agency, so should not be disenfranchised in current developments associated with the 'reinvention' of the modern city encompassed in smart city development. This development will, of course, be affected by who is in charge of shaping the modern city and how they intend to use technology to that end. Technology itself can be limiting in its advancement, accessibility, and acquisition, being restricted byinnovation and the milieu of its development that can restrict production and consumption, including the ability of people to operate it—as for instance in the case of computer software. A balance is, therefore, required between hardware and software (technology and human capital) to improve the quality of life for citizens in the smart city. This necessitates a holistic approach, rather than an unintegrated sector-based approach, where system components or subsystems do not communicate with each other (Mattoni et al. 2015); instead, these authors have advocated for an integrated system that operates much like a whole (human) organism.

It has been argued that elites are responsible for smart technologies coming to cities and causing them to function as platforms for the Internet of Things (IoT) through connections with sensors and computers of various 'intelligence', capable of connecting, communicating, and transmitting information through the Internet (Sadowski & Pasquale 2015). These authors caution against an ensuing 'web of surveillance and power' that results from biometric surveillance capabilities contributing to monitoring and automated policing as part of a 'spectrum of control' that guides governance through 'pervasive surveillance and control mechanisms'. This aspect of the emerging smart city will be considered in more detail in the next chapter, and this chapter will address a broader plethora of problems stemming from the technical dimension. In the next section, technology will be considered as a market-based solution in a technical-entrepreneurial approach to understanding the popularisation of the smart city.

5.1 Technology as a Solution

Computer systems are leading the operation of cities, with their commission stemming from the need to reduce energy consumption and emissions (Lombardi et al. 2017). According to these authors, spatial decision support systems (MC-SDSS), for example, need to be retrofitted, and there is a lack of knowledge and evaluation criteria needed to assess and deliver urban energy using this tool as part of a long-term socioeconomic-environmental approach. Nevertheless, smart technologies are being increasingly deployed in cities for various reasons, including for urban infrastructural control through the integration of urban services with information technology (Luque-Ayala & Marvin 2016). Circulatory flow is managed through networks, such as Rio de Janeiro's Operations Centre (COR)—a control-room scenario (media platform) that has emerged since 2011 to provide logistics at the city-scale from the everyday to emergency situations, such as the traumatic rainfall and flooding experienced in April 2010 that led to the enlistment of IBM to deal with the problem through COR, which operates 24 hours a day and seven days a week and interconnects the information of several municipal systems for visualisation, monitoring, analysis, and response in real-time (Luque-Ayala & Marvin 2016). These authors have defended the 'urban governmentality' that COR represents in addition to offering the novelty of urban vision and engagement.

Wireless technology has now developed well and beyond wired CCTV cameras emplaced to enhance surveillance and, thereby, security, with entire wireless sensor networks capable of (low-power) remote sensing and monitoring a variety of dimensions in the smart city (Ramirez et al. 2016). Data acquired through sensing are stored in compact devices that, according to these authors, do not consume much power and can greatly improve data management in terms of both storage and transmission. In this way, different information can be gathered on various aspects of the environment (and natural hazards), but also accidents and transport, logistics, and healthcare as well as security. Such ICT-led transformations are influencing contemporary responses to global environmental change. As also mentioned by others, such as Sadowski and Pasquale (2015), the role of 'technocratic elites' and that of private capital investing in boosting a techno-environmental fix are recognised, which is part of a wider politico-economic context, so that elites can act to prevent alternative politico-ecological transitions from taking place.

Even though technology, and ICT or information and communication technology in particular, represents a technical approach to evolving cities, urban geographers (Wiig & Wyly 2016) and interdisciplinary networks, such as the Smart Cities Innovation Network (Villanueva-Rosales et al. 2015), have contributed towards understanding smart cities and the rationale for them. Geographers, such as Wiig (2015), have examined IBM's Smart Cities Challenge as an example of policymaking in the smart city. The author portrays initiatives as case studies (also see other publications, e.g. Anthopoulos 2017 for 10 smart city cases)—an approach also adopted in this brief, deployed by various smart city initiatives. He has also addresses the role of city governments as key actors in a multi-stakeholder arena of players responsible for the advancement of the smart city paradigm. Wiig (2015) identifies entrepreneurial governance involved in policy mobility in part of the globalised economy (what he terms as a 'globalised business enterprise' that has attracted corporations like IBM) and digital governance as part of redevelopments to realise the smart city. In a subsequent publication, Wiig (2016) presents the technological solutions provided by the case study of the Digital On-Ramps initiative based on IBM's policy consultation in Philadelphia (also see Wiig 2014), where residents were trained to enter the information and knowledge economy using a workforce education App. He argues that rather than addressing urban inequalities, such programmes work more to sell cities in the global economy. Such a social media style approach to training can become commonplace in the green economy that is still struggling to emerge.

Also relating to policymaking and governance, authors (e.g. Zotano & Bersini 2017) relay opportunities involving Open Data accessible by businesses as well as citizens. According to the authors, Open Data portals can be deployed to develop new business models as part of a holistic approach that they have applied to the Brussels Capital Region. These authors have found that cities, such as Brussels (Belgium), are not fully capable of exploiting the 'real intelligence' provided by smart cities and that the 'maturity' required to achieve this ambition may be attained in the coming years through the implementation of smart city strategies, such as Brussels' Smart City Strategy.

Smart cities can be seen as a contagion that once expelled into cities cannot be reverted and undone. In other words, there is potentially no going back from the smart city craze that has inflicted cities around the world. Should technology be implicit in all, as evidenced by Chourabi et al.'s (2012) smart city initiative framework (see their Fig. 1, p 2294), that recognises two levels of influences: outer factors (natural environment, infrastructure, economy, governance, people, communities) and inner factors (technology, policy, management) that are more influential than the outer factors. The authors consider technology as a 'meta-factor' in smart city initiatives, as it greatly sways all of the other success factors in the framework.

What drives technology, of course, is energy, which is also deserving of consideration, as with smart energy cities (presented in Chap. 6)—a concept that has developed in the literature at least since the early 2010s and is arguably rooted in a sustainability framework. A case-in-point is Milano, Italy as a smart energy city. Smart cities appeared in Italy after 2008, with particular preparation being made heading into Expo 2015. This involved multiple sectors: buildings (domestic, heating); lighting (public, private); transport (public, private); energy use; and energy sources: electricity, natural gas, fuel oil, gasoline, and thermal fluid. Evaluations of energy smartness have been constrained by low-data quality and the availability of energy flows in cities (Causone et al. 2017). The initiative sharing cities accelerated the take-up of smart city solutions; it identified three business models that proved the acceleration of uptake (e.g. refurbishment, smart lamp posts), which was part of doing more with less: smart cities for the age of austerity (Pollio 2016) as part of a technological solution that was supposed to adapt to annihilated fiscal budgets.

Another example of an actually existing smart city is Barcelona, Spain, which has been imagined as a smart and self-sufficient city (smart transformation). Barcelona City Council merged the planning and infrastructure, housing, environment, and ICT departments into a single department called 'Urban Habitat'. A new urban model adopted the vision of Barcelona's chief architect, Vicente Guallart (during the Euro Crisis of 2011–2012) involving the notion of the 'multi-scalar city' as a distributed network, with a vision of empowering citizens through technological improvements. According to March and Ribera-Fumaz (2016), its architecture operates much as a model of networked habitats.

In Portugal, part of connected urban development (CDU) is a leading initiative with CISCO that aims to demonstrate how to leverage ICT above all high connectivity and collaboration. Part of the Portuguese National Plan of Action for Energy Efficiency and National Strategy for Energy, one of these programmes (ECO.AP 2011), aims for an increase of 20% in energy efficiency in public buildings in Leiria, Portugal by 2020.

These cities (Lisbon, also San Francisco, Amsterdam, Seoul, Birmingham, Hamburg, Madrid) will spearhead the implementation of projects aimed at reducing urban emissions of carbon dioxide (CO_2), subsequently acting as references for the widespread implementation of such projects in other cities around the world (CISCO 2008, European Commission 2011; see Galvão et al. 2017). This has been one of the predominant approaches in the emergence of smart cities, which has included the following two main approaches:

- Environmental: sustainable cities, 'green' economy, including energy-efficient buildings; smart mobility—part of a multifaceted, interdisciplinary approach
- Economics: entrepreneurialism, where the business model is used to account for vendors and smart development

The latter encompasses an entrepreneurial approach to digital spaces and Big Data embodies 'spaces of accumulation' that represent commodified digital information.

In a post-capitalist urban and neoliberal context, profit generation is at the forefront of many initiatives building up the notion of smart cities. This is evident through continued efforts since around the time of the European economic crisis and previous to this at a global scale. Technological solutions, although they may not resolve contemporary economic problems, work to support technical groups, as in computing, corporations, and elites that ultimately benefit from this type of urban rebranding and regeneration.

5.2 Social Issues

Although governments have been supporting advancements towards the smart city, there are social issues needing address that provide caveats to such a technological approach. Smart cities are being set up to gather information that can be used to inform decision-making, policymaking, and management. This information is necessary for officials who need to make sensible decisions, as in evidence-based decision-making; in addition, the devices used to collect information have many benefits in that they can be low-energy and their use lead towards energy saving. These platforms have provided an organisation that could even lead to urban economic renewal. As part of an economic development policy, smart cities have been supporting innovation and even included participatory innovation platforms (Anttiroiko 2016). This author, for instance, has written concerning enabler-driven innovation platforms and living labs (e.g. Bates & Friday 2017 based on IoT) that are apparent in Finnish cities, such as Helsinki, Tampere, and Oulu. Such platforms are deployed to support urban revitalisation and economic development even when operating at the level of local governance, where they have stimulated public engagement in the production of local public services and participation in the making of cities. According to the author, participatory innovation platforms help to procure social inclusion, among other things, through platform-based citizen engagement, which is considered to be a 'soft' strategy to counteract social polarisation and socioeconomic segregation and, therefore, inequalities.

Public engagement with smart cities and their growth is evident in various forms. One piece of evidence are the publications that have proliferated recently, as for example special issues addressing smart city technology (e.g. He et al. 2014) and sustainable urban transformation (Zhang et al. 2016) as well as the aforementioned special issue by Wiig and Wyly (2016)—based on an Association of American Geographers meeting that addressed the question: What does the smart city, as a digital turn in urban governance, tell us about cities today? that acknowledged the transformative process demonstrated by smart cities—plus the special section on rapid urbanisation by Wigginton et al. (2016), to name a few. Another example is that of university training courses based on an innovative learning system in entrepreneurship using mass open online courses to support policy learning (Holotescu et al. 2016). As already mentioned, an entrepreneurial approach is evidenced in smart cities, with markets recognised—as for example hydrogen as an electric carrier and for storage over electric batteries (Marino et al. 2015); additionally, Sadowski (2016) recognises the need to 'sell smartness' and, by so doing, conveys its commodification where there is wealth in cities. In fact, niche markets are apparent, engaging all business sectors (and multi-stakeholders) and headed by local governments in conjunction with vendors (Anthopoulos & Fitsilis 2015). So, in addition to the predominantly economic driver of smart cities, there are also social systems of consumership that are both affected by as well as driving change where there is wealth. Renewable energy, such as solar energy and PV (Menniti et al. 2017), has been advocated to fuel sustainable and smart cities (e.g. Barragán & Terrados 2017), conveying an environmental approach guiding their development (e.g. Katra town, India; Sharma & Dogra 2017). Such 'urban entrepreneurialism', which is part of the corporate smart city model, allows for urban competitiveness, driven by hi-tech companies and city governance (referred to by Hollands 2015) as corporate and entrepreneurial governance, but according to the author constrains public participation in the smart city.

Among this growing body of data are the issues of data mass capture and surveillance that emerge with spatial data that are possibly tracked by digital technology (Cho 2017). Digital data can also relay anytime-anywhere information, providing access and control over people's movements (Cosgrave et al. 2013). These authors also mention 'information marketplaces' that point to the potential for commodified information, with implications for national private security. In addition to issues of hackers and accessibility issues, there are also robotics to consider in keeping security. As for example witnessed by Odendaal (2006), who recognises the socioeconomic fragmentation of South African cities and the potential for manipulation by corporations, such as the South Africa company Desert Wolf that discharged the Skunk—a riot-control drone armed with sublethal capabilities (fires paintballs, pepper-spray, rubber bullets, blinding lasers)—to disperse or mark people in crowds, such as protestors (Doctorow 2014). According to the author, this technology, used by mining companies against strikes in South Africa, has potential to be deployed to subdue those who seek to interrupt and change the current structures of power and capital. So, there are other social issues that are mixed up with technological approaches to security as well as other aspects of smart cities.

By implication, more research is needed to address how smart cities fit into a democratic society. Democratic governance is counterposed by elitism and potentially automated processes (e.g. e-government) and such top-down organisations that have potential to police human behaviour. The amassment and use of Big Data, including as for example Big Data analytics (e.g. Al Nuaimi et al. 2015), pose a challenge to privacy due to a lack of public consent. This could act to sharpen the private-public boundary, as by recognising that by stepping outside one's house is stepping into the monitored, public domain. However, through smart houses themselves, human behaviour can be monitored even within the private sphere (e.g, smart home monitoring systems), so that the notion of privacy is once again superseded by constant observation, monitoring, and potentially control.

References

Al Nuaimi E, Al Neyadi H, Mohamed N, Al-Jaroodi J (2015) Applications of big data to smart cities. J Internet Ser Appl 6(25):15 p. https://doi.org/10.1186/s13174-015-0041-5

Anthopoulos L (2017) Smart utopia vs smart reality: learning by experience from 10 smart city cases. Cities 63:128–148. https://dx.doi.org/10.1016/j.cities.2016.10.005

Anthopoulos LG, Fitsilis P (2015) Understanding smart city business models: a comparison. WWW'15 companion, Florence, Italy, 18–22 May 2015, pp 529–533. https://dx.doi.org/10.1145/2740908.2743908

Anttiroiko A-V (2016) City-as-a-platform: the rise of participatory innovation platforms in Finnish cities. Sustainability 8(9):922 (31 p). https://doi.org/10.3390/su8090922

Barragán A, Terrados J (2017) Sustainable cities: an analysis of the contribution made by renewable energy under the umbrella of urban metabolism. Int J Sus Dev Plann 12(3):416–424. https://doi.org/10.2495/SDP-V12-N3-416-424

Bates O, Friday A (2017) Beyond data in the smart city: repurposing existing campus IoT. IEEE Pervasive Comput 16(2):54–60. https://doi.org/10.1109/MPRV.2017.30

Causone F, Sangelli A, Pagliano L, Carlucci S (2017) An exergy analysis for Milano smart city. Energy Proced 111:867–876. https://doi.org/10.1016/j.egypro.2017.03.249

Cho L (2017) Mass capture: the making of non-citizens and the mainland travel permit for Hong Kong and Macau residents. Mobilities 12(2):188–198. https://doi.org/10.1080/17450101.2017.1292776

Chourabi H, Nam T, Walker S, Gil-Garcia JR, Mellouli S, Nahori K, Pardo TA, Scholl HJ (2012) IEEE Xplore, 09 Feb 2012, pp 2289–2297. https://dx.doi.org/10.1109/HICSS.2012.615

Cosgrave E, Arbuthnot K, Tryfonas T (2013) Living labs, innovation districts and information marketplaces: a systems approach for smart cities. Procedia Comput Sci 16:668–677. https://doi.org/10.1016/j.procs.2013.01.070

Doctorow C (2014) Riot control drone that fires paintballs, pepper-spray and rubber bullets at protesters. Boing Boing, 17 June 2014. https://boingboing.net/2014/06/17/riotcontroldronethatpaintb. Accessed 26 June 2015

Galvão JR, Moreira L, Gaspar G, Vindeirinho S, Leitão S (2017) Energy system retrofit in a public services building. Manage Environ Qual 28(3):302–314. https://doi.org/10.1108/MEQ-02-2014-0028

He Y, Stojmenovic I, Liu Y, Gu Y (2014) Smart city. Int J Distrib Sens Net 10(5):2. https://doi.org/10.1155/2014/867593

Hollands RG (2015) Critical interventions into the corporate smart city. Camb J Reg Econ Soc 8:61–77. https://doi.org/10.1093/cjres/rsu011

Holotescu C, Slavici T, Cismariu L, Gotiu LOL, Grosseck G, Andone D (2016) MOOCs for innovative entrepreneurship in smart cities. World J Educ Technol 8(3):245–251. https://doi.org/10.18844/wjet.v8i3.832

Lombardi P, Abastante F, Moghadam ST, Toniolo J (2017) Multicriteria spatial decision support systems for future urban energy retrofitting scenarios. Sustainability 9:1252 (pp 14). https://dx.doi.org/10.3390/su9071252

Luque-Ayala A, Marvin S (2016) The maintenance of urban circulation: an operational logic of infrastructural control. Environ Plann D 34(2):191–208. https://doi.org/10.1177/0263775815611422

March H, Ribera-Fumaz R (2016) Smart contradictions: the politics of making Barcelona a self-sufficient city. Eur Urban Reg Stud 23(4):816–830. https://doi.org/10.1177/0969776414554488

Marino C, Nucara A, Pietrafesa M (2015) Electrolytic hydrogen production from renewable source, storage and reconversion in fuel cells: the system of the "Mediterranea" University of Reggio Calabria. Energy Proced 78:818–823. https://doi.org/10.1016/j.egypro.2015.11.001

Mattoni B, Gugliermetti F, Bisegna F (2015) A multilevel method to assess and design the renovation and integration of smart cities. Sustain Cities Soc 15:105–119. https://doi.org/10.1016/j.scs.2014.12.002

Menniti D, Bayod-Rújula AA, Burgio A, García DAL, Leonowicz Z (2017) Solar energy and PV systems in smart cities. Int J Photoenergy 3574859, p 2. https://doi.org/10.1155/2017/3574859

Odendaal N (2006) Towards the digital city in South Africa: issues and constraints. J Urban Technol 13(3):29–48. https://doi.org/10.1080/10630730601145997

Pollio A (2016) Technologies of austerity urbanism: the "smart city" agenda in Italy (2011–2013). Urban Geogr 37(4):514–534. https://doi.org/10.1080/02723638.2015.1118991

Ramírez CA, Barragán RC, García-Torales G, Larios VM (2016) Low-power device for wireless sensor network for smart cities. IEEE Xplore, 13 Feb 2017, p 3. https://dx.doi.org/10.1109/LAMC.2016.7851298

Rose G (2017) Posthuman agency in the digitally mediated city: exteriorization, individuation, reinvention. Ann Am Assoc Geogr 107(4):779–793. https://doi.org/10.1080/24694452.2016.1270195

Sadowski J (2016) Selling smartness: visions and politics of the smart city. Doctoral thesis, Arizona State University, p 223 https://repository.asu.edu/items/40245

Sadowski J, Pasquale F (2015) The spectrum of control: a social theory of the smart city. University of Maryland Francis King Carey School of Law, Legal studies research paper no. 2015–26. https://ssrn.com/abstract=2653860

Sharma AK, Dogra VK (2017) Preparation of papers—potential alternate energy resources for sustainability: a must need for a top pilgrimage city. Energy Proced 115:173–182. https://doi.org/10.1016/j.egypro.2017.05.047

Thornbush M, Golubchikov O, Bouzarovski S (2013) Sustainable cities targeted by combined mitigation–adaptation efforts for future-proofing. Sustain Cities Soc 9:1–9. https://doi.org/10.1016/j.scs.2013.01.003

Villanueva-Rosales N, Cheu RL, Gates A, Rivera N, Mondragon O, Cabrera S, Ferregut C, Carrasco C, Nazarian S, Taboada H, Larios VM, Barbosa-Santillan L, Svitek M, Pribyl O, Horak T, Procazkova D (2015) A collaborative, interdisciplinary initiative for a smart cities innovation network. IEEE Xplore, 28 Dec 2015, pp 1–2. https://dx.doi.org/10.1109/ISC2.2015.7366179

Wiig A (2014) After the smart city: global ambitions and urban policymaking in Philadelphia. Doctoral thesis, Temple University, p 326. https://digital.library.temple.edu/cdm/ref/collection/p245801coll10/id/294272

Wiig A (2015) IBM's smart city as techno-utopian policy mobility. City 19(2–3):23. https://doi.org/10.1080/13604813.2015.1016275

Wiig A (2016) The empty rhetoric of the smart city: from digital inclusion to economic promotion in Philadelphia. Urban Geogr 37(4):535–553. https://doi.org/10.1080/02723638.2015.1065686

Wiig A, Wyly E (2016) Introduction: thinking through the politics of the smart city. Urban Geogr 37(4):485–493. https://doi.org/10.1080/02723638.2016.1178479

Wigginton NS, Fahrenkamp-Uppenbrink J, Wible B, Malakoff D (2016) Cities are the future: rapid urbanization is overtaxing the planet, but it may not have to. Science 352(6288):904–905. https://doi.org/10.1126/science.352.6288.904

Zhang X, Hes D, Wu Y, Hafkamp W, Lu W, Bayulken B, Schnitzer H, Li F (2016) Catalyzing sustainable urban transformations towards smarter, healthier cities through urban ecological infrastructure, regenerative development, eco towns and regional prosperity. J Clean Prod 122:2–4. https://doi.org/10.1016/j.jclepro.2016.02.038

Zotano MAG, Bersini H (2017) A data-driven approach to assess the potential of Smart Cities: the case of open data for Brussels Capital Region. Energy Proced 111:750–758. https://doi.org/10.1016/j.egypro.2017.03.237

Chapter 6
Conclusion

Abstract In this final chapter, the problems posed by smart development are considered from an ethical perspective. Here, the business model and entrepreneurial smart growth are examined for implications to unchecked development based on artificial intelligence (AI) as being centrally-controlled by computer technicians and corporations. Caveats are presented for consideration of potential developments stemming from information technology or IT corporations and their involvement in the expansion of robotics, including those directly involved in smart cities.

Keywords Integrated (holistic) systems · Interdisciplinary approach · Framework/taxonomies for classification · Models · Robotics · Smart control · Democratic governance · Urbanisation · Sustainable development · International ethics

Cities around the world have been getting 'smarter' as more advanced technology is integrated into urban planning and design. People are relying more on technology for routine communications at home and in their daily lives, which has merited use of information and communication technology (ICT) with unified communications, such as telecommunications (including smartphones, telephone lines, wireless networks, etc.) as well as computers, iPads, going paperless, chips and batteries for energy storage, wireless/remote charging, and so on. Cities are also adopting new technology through sensors deployed to monitor and gather information about people and their environment for reasons of ease-of-use and resource efficiency, entailing 'doing things better for less' or more for less as part of a descaling economy and austerity measures. Such increasing automation and efficiency have led to Big Data collection that is used for reporting and governance and loops back to inform planning and design.

This brief has examined the progression of the development of the smart city and its dialogue with the concept of sustainability, from its infancy at the scale of individual buildings mixed with the literature on efficient buildings and moving towards well-adapted and resilient sustainable cities in transition. More recently, there has been a focus in the literature on energy savings and efficiency that has fostered research progress in intelligent and smarter systems. These have evolved more recently to embrace the concept of the smart energy city. Specifically, this brief examines the recent evolution of the application of smart cities as 'actually existing smart cities' that is driving these cities as they emerge around the world.

By addressing 'actually existing' smart cities, it is possible to exclude from the analysis of utopian approaches that are non-existing. Tracking the smart cities of today, it is possible to illuminate the progress made in the recent past, in the present, and consider the culmination of the current trajectory. Amid the plethora of developing smart cities and the attention paid to them through various publications, reports, and conferences evident especially since 2016 is a need to consider the direction that smart development has actually taken in contemporary cities. Ultimately, this will influence future development that is necessarily shaped by current real-world experiences, including the resolution of any challenges.

The contributions of this brief have included a delineation of the development of smart cities that has, arguably, stemmed very much from energy-efficient cities and low carbon urbanism. It has also adopted a sociotechnical approach to examine the evolution of the concept and its continuation towards the 'smart energy city'. This line of progression, which is grounded in conceptualisations of sustainable cities, has emerged to once again encompass energy smartness—and has, thereby, gone full circle. The role that technology has had in instigating and developing smart cities cannot be overstated. It has been fundamental to controlling climate in buildings and improving mobility and other aspects of urban development, including crowdsourcing based on the amassment of Big Data.

There are technical and social issues outlined in this brief, such as the implications of control extending to people as well as technology, as for instance using machines such as the Skunk in South Africa for controlling protests and to disperse crowds. Due to piecemeal smart development, the issue of social justice may be overlooked, and attention is needed here as part of democratic governance, which can continue to be disregarded if entrepreneurial governance goes unchecked, as apparent already with some of the mining corporations of South Africa. Poverty is another social issue needing continued attention by the government, in particular, to offset growing disparities in expanding smart cities. There are contingent issues here, such as that of urbanisation, that similarly affect smart city development. Disregarding the issues will not make them go away; instead, they will continue to propagate in the actually existing smart cities of today and tomorrow.

Control through technology, such as that of monitoring and information gathering, will not on its own solve the wicked problems facing society. Big Data can be generated, for instance, but on its own problems will persist—what society does with the information that it collects is vital, and will determine how problems are addressed and resolved. This information can be disseminated to corporations for marketing purposes and lead to unethical use without consent, as already cautioned for the waterfront smart development planned for the City of Toronto (Canada). It may also venture into the private domain (people's homes) and there gather knowledge about individuals and their livelihoods, affecting personal security (privacy) even in the private domain. Again, how this information is used can interfere with people's rights and notions of democracy.

Moving from the control of the technical to that of the social (and individual) is a key consideration, especially for countries that are already governed by dictators or those ruled by technocratic elites (as discussed by Sadowski & Pasquale 2015; cf. Brenner & Theodore 2002) in so-called democratic countries. Neoliberalism and the persistence of capitalism are also affecting smart development through entrepreneurial ventures involved in selling devices as well as the idea of utopian futuristic cities. Vendors are able to sell information technology or IT in various forms to build the city of tomorrow as a technologically advanced entity that benefits them, but without planning for the challenges that such development introduces.

Social responsibility is needed to be upheld by corporations as well as individual sellers of these future cities. They should represent more than a contribution to sales and profit margins in the business model. It should be the imperative of leaders and governments, in particular, to closely monitor and check smart growth as part of a strategy grasped around the world. What is more, an autonomous ethical governing body is needed to examine the direction that these cities are taking as part of sustainable development and, more specifically, the sustainable development goals (SDGs) for sustainable cities and communities.

Smart development, after all, is directed—the question is by whom and to what end. This needs to be examined from a community-based framework steered towards the common good. People and organisations with vested interests need to be governed and their actions ethically verified before smart growth persists in developed and developing nations. Robotics, likewise, needs to be closely monitored and controlled before the situation is reversed and people themselves become the target of monitoring and control by intelligent technology and the corporations that deliver these entities into existence. It is important to do so in order to avoid a future where human ethics and freedoms are driven to the brink to be replaced by artificial intelligence (AI, robotics, e.g. Sophia developed by Hanson Robotics) that can become self-conscious and self-directed entities and potentially in control of society.

The United Nations has the potential to develop an international ethics committee that is responsible for upholding conventional democracy and the personal and individual freedoms of all humans. What seems like an innocent development in smartness and intelligent (automated) operations can turn into a social nightmare for people around the world. Governmental organisations need to acknowledge their role in curbing any potential threat to democracy, even in the face of elites and other undemocratic groups, that are increasingly empowered around the world and taking advantage of such a non-thinking mechanism as independent devices that are centrally-controlled via the Internet to operate within a platform that can be hacked or controlled by IT-savvy technicians who can be bought.

The cities of tomorrow, and the operation of society that is increasingly contained in these cities, rest on computer security and IT control that eludes many people, including the decision-makers of today. It is the role of computer-literate professionals and governments to ensure the safety of these places and their continued commitment to the individual rights and freedoms on which democracy is based.

Effective checks and balances need to be installed and in place for healthy smart development. Reporting needs to occur to officials who are in charge of these controls that govern smart cities and their propagation from the city to city-region to envelope entire countries and eventually the world.

References

Brenner N, Theodore N (2002) Cities and the geographies of 'actually existing neoliberalism'. Antipode 34:349–379. https://doi.org/10.1111/1467-8330.00246
Sadowski J, Pasquale F (2015) The spectrum of control: a social theory of the smart city. University of Maryland Francis King Carey School of Law, Legal studies research paper no. 2015–26. https://ssrn.com/abstract=2653860

Index

© The Author(s), under exclusive license to Springer Nature Switzerland AG 2020   63
M. J. Thornbush and O. Golubchikov, *Sustainable Urbanism*
in Digital Transitions, SpringerBriefs in Geography,
https://doi.org/10.1007/978-3-030-25947-1